4th International Conference on Mechanical, System and Control Engineering 2020 (ICMSC 2020)

IOP Conference Series: Materials Science and Engineering
Volume 904

Moscow, Russia
20 - 23 June 2020

ISBN: 978-1-7138-2814-3
ISSN: 1757-8981

Printed from e-media with permission by:

Curran Associates, Inc.
57 Morehouse Lane
Red Hook, NY 12571

Some format issues inherent in the e-media version may also appear in this print version.

Institute of Physics
Dirac House, Temple Back
Bristol BS1 6BE UK

Phone: 44 1 17 929 7481
Fax: 44 1 17 920 0979

techtracking@iop.org

Additional copies of this publication are available from:

Curran Associates, Inc.
57 Morehouse Lane
Red Hook, NY 12571 USA
Phone: 845-758-0400
Fax: 845-758-2633
Email: curran@proceedings.com
Web: www.proceedings.com

TABLE OF CONTENTS

IOP Conf. Series: Materials Science and Engineering **904** (2020) 011001 doi:10.1088/1757-899X/904/1/011001

PREFACE

The wonderful conference, the 4th International Conference on Mechanical, System and Control Engineering (ICMSC 2020), which should be taken place in Russia traditionally since 2017, was held virtually under the unusual conditions that COVID-19 virus imposes this year. The safety and well-being of all conference participants is our priority. The COVID-19 is unpredictable, so conference postponement met uncertainty, while many scholars and researchers want to attend this long-waited conference and have academic exchanges with their peers.

2020 is certainly not easy - the situation with the COVID-19 virus is not easy in the whole world. But I want to note that there are no barriers to science, and we continue to work on our research areas remotely, using modern technical means. ICMSC 2020 chose Zoom as technical support to hold this virtual conference and each presenter holds 15 minutes for their work and have a pleasant discussion after presentation. Before conference, all the authors have prepared a video as backup in case there might be some technical or network problem, eventually, ICMSC 2020 gain a great success and became an effective communication platform for all the participants.

The subject of the conference is devoted to the mechanics, systems and control, and has many applications in science and technology. Traditionally, the conference reviewed reports on the engineering industry, production organization and the space industry, and brought together participants from many countries from Russia, China and there were representatives of European countries. They communicated scientific sections were at the conference: Control Theory and Engineering, System Modelling, Power Engineering and simulations, etc. They put up various scientific approaches to solving similar problems.

This is the most important advantage of an international conference to exchange experiences. I believe that the experience that we get from mutual communication using technical means will help not stop the progress in the development of engineering and engineering science.

Finally, I wish you healthy, friends and every success!

Conference Steering Chair

Vsevolod V. Koryanov

Bauman Moscow State Technical University, Russia

IOP Conf. Series: Materials Science and Engineering **904** (2020) 011002 doi:10.1088/1757-899X/904/1/011002

Scientific Advisory Chairs

Prof. Yuriy F. Gortyshov, President of Kazan National Technical University - KAI, Russia
Prof. Victor Kazakovtsev, Bauman Moscow State Technical University, Russia

Conference Chairs

Prof. Marina Polyakova, Nosov Magnitogorsk State Technical University, Russia
Prof. Yury Klochkov, Peter the Great St. Petersburg Polytechnic University, Russia

Steering Chairs

Prof. Lyudmila K. Kuzmina, Kazan National Technical University - KAI, Russia (Head of IFNA RC , Member of National Russian Committee on TPM)
Prof. Vsevolod V. Koryanov, Bauman Moscow State Technical University, Russia

Program Chairs

Prof. Oleg Gennadievich Morozov, Chair of Department in Kazan National Technical University - KAI, Russia
Assoc. Prof. Evgeny Burnaev, Skolkovo Institute Science and Technology, Russia
Prof. Ricardo Nuno Madeira Soares Branco, Universidade de Coimbra, Portugal
Prof. Antonio Maffei, KTH- Royal Institute of Technology, Sweden
Assoc. Prof. Galder Kortaberria, University of the Basque Country, Spain

Workshop Chair

Prof. Vladimir A. Frolov, Samara State Aerospace University, Russia

Poster Chair

Assoc. Prof. Maxim Bakaev, Novosibirsk State Technical University, Russia

International Committee

Prof. Alexander Kuzmin, St Petersburg State University, Russia
Prof. Alexey Zhdanov, Vladimir State University named after Alexander and Nikolay Stoletovs, Russia
Prof. Alla Brom, Bauman Moscow State Technical University, Russia
Prof. Andrey Ronzhin, SPIIRAS, Russia
Prof. Adrian Olaru, University Politehnica of Bucharest, Romania
Prof. John Watkins, Wichita State University, USA
Prof. Andrzej Gontarz, Lublin University of Technology, Poland
Prof. Boris Alpatov, Ryazan State Radio-Engineering University, Russia
Prof. Constantin Daniel Comeaga, University POLITEHNICA of Bucharest, Romania
Prof. Dmitry Pashchenko, Penza State University, Russia
Prof. Eugene Larkin, Tula State University, Russia

Prof. Evgeni Magid, Innopolis University, Russia
Prof. Faris Tarlochan, Qatar University, Qatar
Prof. Igor Olegovich Temkin, National University of Science and Technology MISiS, Russia
Prof. Irina N. Omelchenko, Bauman Moscow State Technical University, Russia
Prof. Leonid V. Bessonov, Saratov State University, Russia
Prof. Liangen Yang, Hubei University of Technology, China
Prof. Meng Yi, Chongqing University, China
Prof. Nikolay D. Polyakhov, Saint-Petersburg Electrotechnical University LETI, Russia
Prof. Nikolay Karabutov, MIREA - Russian Technological University, Russia
Prof. Pavel A. Drogovoz, Bauman Moscow State Technical University, Russia
Prof. Ruchkin Vladimir Nikolaevich, Ryazan State University named for S.Yesenin, Russia
Prof. Shyh-Chour Huang, National Kaohsiung University of Science and Technology, Taiwan
Prof. Thananchai Leephakpreeda, Thammasat University, Thailand
Prof. Valentin Morozov, Vladimir State University named after Alexander and Nikolay Stoletovs, Russia
Prof. Valeriy Matveev, Samara National Research University, Russia
Prof. Viktor Eremeev, Ryazan State Radio Engineering University, Russia
Prof. Vladimir Belevitin, Chelyabinsk State Pedagogical University, Russia
Prof. Yoshifumi Yokoi, National Defense Academy of Japan, Japan
Prof. Yuri N. Skiba, Universidad Nacional Autónoma de México, Mexico
Prof. Yaguang Zhu, Chang'an University, China
Prof. Gheorghe-Daniel Andreescu, Universitatea Politehnica Timisoara, Romania
Prof. Debao Zhou, University of Minnesota, USA
Prof. Kojiro Iizuka, Shibaura Institution of Technology, Japan
Prof. Jiann-Shiou Yang, University of Minnesota, USA
Prof. Shunming Li, Nanjing University of Aeronautics and Astronautics, China
Prof. Valery Matveev, Samara National Research University, Russia
Prof. Vadim Kramar, Sevastopol State University, Russia
Prof. Alexander Pesin, Nosov Magnitogorsk State Technical University, Russia
Prof. Hanan Anwar, Housing and Building National Research Center, HBRC, Egypt
Prof. Kuncoro Diharjo, Universitas Sebelas Maret, Indonesia
Prof. Pei-Hsing Huang, National Yunlin University of Science and Technology, Taiwan
Prof. Prashant Date, Indian Institute of Technology, Bombay, India
Prof. Sung Ho Yang, Korea National University of Education, Korea
Prof. Timur Atabaev, Nazarbayev University, Kazakhstan
Prof. Yury Ivanov, Bauman Moscow State Technical University, Russia
Assoc. Prof. Giuseppina Gini, Politecnico di Milano, Italy
Assoc. Prof. Rosli Ahmad, Universiti Tun Hussein Onn Malaysia
Assoc. Prof. Ibrahim Sultan, Federation University, Australia
Assoc. Prof. Anastasiya Yu. Sitnikova, Samara State Aerospace University, Russia
Assoc. Prof. Evgeniya Gorlacheva, Bauman Moscow State Technical University, Russia

IOP Conf. Series: Materials Science and Engineering **904** (2020) 011002 doi:10.1088/1757-899X/904/1/011002

IOP Conf. Series: Materials Science and Engineering **904** (2020) 011004 doi:10.1088/1757-899X/904/1/011004

Peer review declaration

All papers published in this volume of *IOP Conference Series: Materials Science and Engineering* have been peer reviewed through processes administered by the Editors. Reviews were conducted by expert referees to the professional and scientific standards expected of a proceedings journal published by IOP Publishing.

- **Type of peer review: Double-blind**
- **Conference submission management system: iConf Conference System**
- **Number of submissions received: 39 submissions**
- **Number of submissions sent for review: 39 submissions**
- **Number of submissions accepted: 23 submissions**
- **Acceptance Rate (Number of Submissions Accepted / Number of Submissions Received X 100): 58.97%**
- **Average number of reviews per paper: 2**
- **Total number of reviewers involved:20 reviewers**
- **Any additional info on review process:**

  Accepted papers should follow:

  Introduce an area that appears promising, or might stimulate others to develop promising alternatives; The paper is clearly written and organization is good. The figures have proper aesthetics. The mathematical equations are given correctly. The conclusions are proper. The references are modern.

- **Contact person for queries:**
  vsevolod.v.koryanov
  vsevolod.v.koryanov@gmail.com

IOP Conf. Series: Materials Science and Engineering **904** (2020) 012001 doi:10.1088/1757-899X/904/1/012001

Pulse laser thermography as a non-destructive method of latent surface defects of materials

Pavel I. Abramov, Eugene V. Kuznetsov, Peter Y. Lobanov, Oleg E. Sidoryuk and Leonid A. Skvortsov[a]

JSC Polyus Research Institute of M.F.Stelmakh, Moscow, Russia

[a]lskvortsov@gmail.com

Abstract. The time dependences of the surface temperature increment of painted metal objects were studied depending on the thickness of the layer subjected to corrosion, as well as putty, restoring erosion defects on it. Hidden under a layer of paint, they cannot be detected remotely during normal visual observation. At the same time they are well detected by pulsed laser thermography. The experimental data in the work are compared with the results of mathematical modeling. Particular attention is paid to eliminating possible distortions caused by external contamination of the tested objects.

1. Introduction

Remote detection of latent defects on the surface of materials based on active photothermal radiometry (thermography) is a very attractive technology, the beginning of which was laid in the late 70s - early 80s of the last century [1 - 4]. Over the years, this technology has undergone significant development. In the broad sense of the word, one can speak of flash thermography [5, 6] and modulation (lock-in) IR thermography [7, 8]. In this case pulsed, or non-stationary, thermography is the most common form of non-destructive assessment of the object of study using the thermographic method (NDE, nondestructive evaluation) [9 - 13]. This is due to the possibility of studying significant areas on the surface of the object in the presence of a high-speed IR camera with a large field of view, which is of fundamental importance for increasing the productivity of non-destructive testing of materials. Therefore, the main attention in the future will be given to the possibilities and development of this particular type of material research.

The physical principle underlying the use of flash thermography for non-destructive testing is to measure the time dependence of surface temperature changes, while lock-in thermography allows you to study the phase of the photothermal signal. In both cases, the difference in thermal diffusion in different places of the sample due to the presence of defects and inhomogeneities in it affects the measurement result. In other words, subsurface defects change the rate of thermal diffusion, thereby manifesting itself in a change in the parameters of the recorded photothermal signal in the places where defects exist relative to areas where they are absent. Any inhomogeneities, such as inclusions, subsurface defects, or porous formations distort the propagation of thermal waves and modify the temperature response on the surface of the sample in comparison with the defect-free region. Due to the direct relationship between the thermal and mechanical properties, the detection of the distribution of the thermal signal allows mapping of the physical properties of the material under study during its scanning.

IOP Conf. Series: Materials Science and Engineering **904** (2020) 012001 doi:10.1088/1757-899X/904/1/012001

The development of flash thermography and lock-in thermography methods is pulse phase thermography based on the application of the Fourier transform to the function of temperature change over time [14, 15]. Pulse phase thermography maximally implements the idea of taking into account the fundamental differences in temperature evolution in defective and defect-free (reference) areas of the test object. However, its implementation in practice requires considerable time, since one has to deal with sequences consisting of several hundred and even thousands of images.

In the classic active pulsed thermographic technique (flash thermography), the studied surface of the object is heated under the influence of radiation from one or several flash lamps, after which its cooling is recorded using an IR camera. This methòd has long been used for non-destructive and non-contact investigation of interfaces between the surface layer (coating) and the base material (substrate), for example, to detect adhesion defects (delaminations), local corrosion foci, surface cracks that change the one-dimensional heat flux from the surface [13].

However, a significant divergence of radiation from flash lamps and its low spectral brightness significantly limit the scope of flash thermography, in particular, when testing objects located at a considerable distance from the heating source. At the same time, just such a task is currently gaining more urgency and attention of researchers, since it is related to remote monitoring of structures, for example, in the form of bridges, tanks for transporting and storing petroleum products, liquefied gas, car bodies, building walls and etc.

The aim of this work is to demonstrate a new approach for remote monitoring of the surface condition of various structures located at considerable distances from the heating source using pulsed laser thermography, which excludes the influence of external factors (for example, contamination of the object's surface) on the detection of "camouflaged" subsurface defects.

The solution to this problem became possible in connection with the creation of powerful and at the same time compact semiconductor lasers emitting at different wavelengths of the electromagnetic spectrum (from UV to far IR).

2. Experimental setup

Fig.1 shows an experimental setup for remote analysis of the surface of materials by means of pulsed laser thermography.

The studied sample (1) is affected by laser radiation pulses (2) formed by an electromechanical chopper (3).

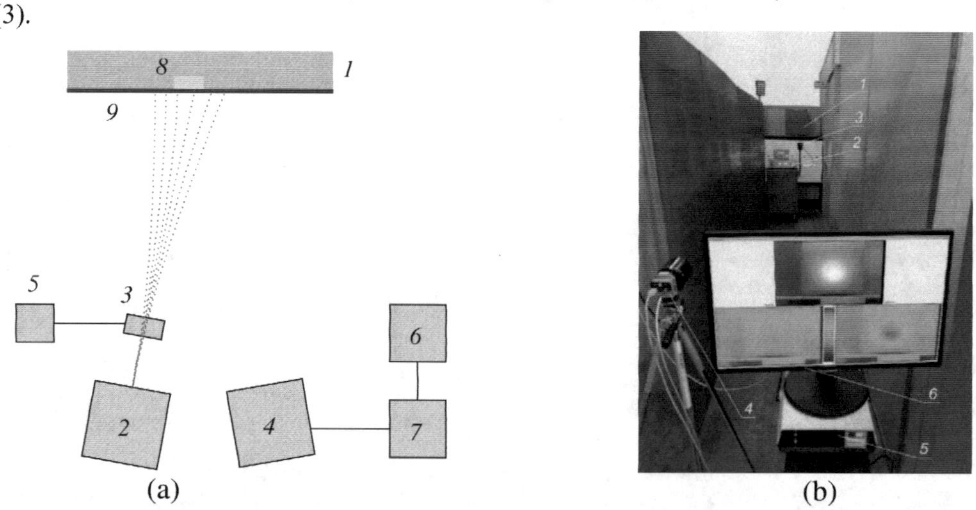

(a) (b)

Figure 1. a) Scheme of non-destructive testing of hidden subsurface defects using laser pulse thermography; b) Photo of the experimental setup.

1 — sample, 2 — laser, 3 — radiation chopper, 4 — infrared camera, 5 — control panel, 6 — monitor, 7 — computer, 8 — hidden subsurface defect, 9 — protective coating (paint)

The source of laser radiation in the work was the assembly of semiconductor laser diodes (average radiation power 90 W) with a generation wavelength of 915 nm, having a common optical fiber output. The duration of laser exposure was set by a timer from the control panel (5), which was connected electrically with a laser chopper. An IR camera (4) was used to register a map of local temperature increments on the surface of the test sample. Thermographic images were presented on a monitor screen (6) and could be adjusted using a special computer program (7).

The laser exposure area and the corresponding power of laser radiation density could be changed using a tunable lens.

Samples in the form of bars of structural steel of the St3 grade with surface defects (8) under a dark protective coating layer (9) were used as a target. At the same time, the areas exposed to corrosion on the surface of the test areas did not appear in any way when remotely observing samples using a visible range CCD camera or a thermal imaging camera in the absence of laser radiation.
As an infrared camera, FLIR Tau2 with a resolution of 336 × 256 with a pixel size of 17 μm and a sensitivity of 0.05 mK was used.

3. Thermographic results

When a sample is irradiated with laser radiation, latent defects in the form of rust-affected areas are clearly visible when observed using an infrared camera located at a distance of 10 m associated with a laser system (Fig. 2). The same figure shows the result of visualization of a hidden defect under similar conditions in the form of cross-shaped surface cracks 2 mm deep, obtained by means of spark erosion, coated with polyester putty, followed by polishing and painting the surface with alkyd enamel. Such masking of surface cracks also made it impossible to visualize them in the visible range of the spectrum even at high magnification, as well as when observed with a thermal imager. However hidden defects of this kind can be visualized using pulsed laser thermography (Fig. 2b).

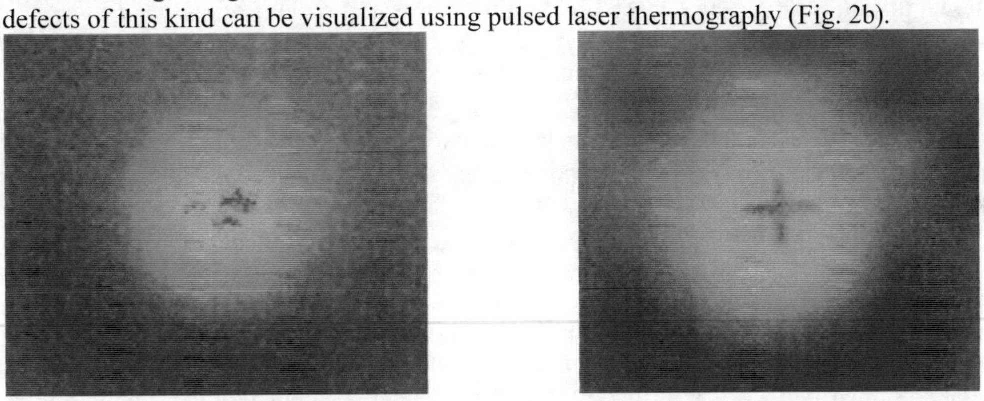

(a) (b)

Figure 2. Thermographic images observed after laser irradiation of a metal plate with local defects under a layer of paint: (a) iron corrosion centers, (b) cross-shaped surface cracks

In the above figures a thermal footprint is clearly visible due to a violation of the heat sink in the defective area of the test sample, hidden by a protective coating. In this case, the paint coating absorbed at least 80% of the radiation energy, the emissivity in the region of 7.5 - 13 μm was 0.9.

Fig.3 shows the time dependences of the increment in the surface temperature of the sample depending on the thickness of the layer subjected to corrosion (Fig. 3a) and the thickness of the protective layer hiding cross-shaped traces of erosion (Fig. 3b).

These dependencies can be used as the basis for quantitative estimates of the parameters of latent surface defects in materials.

Under real conditions, the detection of hidden near-surface defects by pulsed laser thermography can distort the presence of a number of factors. In particular, external surface contamination can give an ambiguous interpretation in the infrared region. The presence on the surface of areas with increased absorption of laser radiation leads to their overheating in comparison with the rest of the sample. On the other hand, the presence of stains of a foreign substance can be accompanied by a local change in the emissivity of the surface. In both cases, testing the object by means of laser thermography gives a picture of heterogeneity, which can hide the presence of defects under the surface or distort their nature.

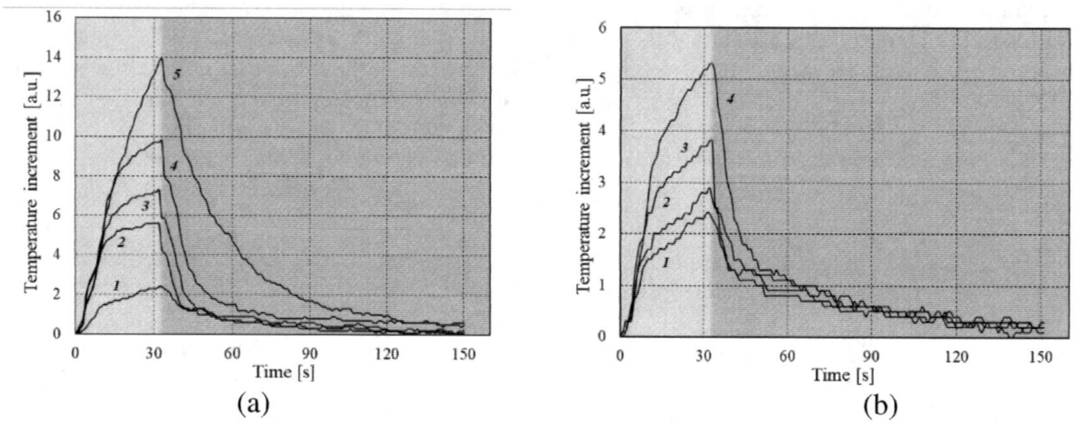

Figure 3. Time dependences of changes in the surface temperature of plates made of St3 steel when exposed to laser radiation with a power density of 0.4 W/cm^2 for 33 seconds and subsequent cooling for different thicknesses h of the surface layers of (a) corrosion products of hydrated iron oxide: 1- h = 0,25 mm, 2- h = 0,5 mm, 3- h = 0,8 mm, 4- h = 1,1 mm, 5- h = 2,3 mm; (b) polyester putty: 1- h = 0,25 mm, 2- h = 0,5 mm, 3- h = 1,6 mm, 4- h = 2,5 mm

It is possible to eliminate false responses by comparing the images of the thermal imager with the corresponding images of the visible range camera, on which obvious surface contamination is usually clearly observed. And only when a homogeneous picture is recorded in the latter case, we can talk about the possibility of detecting latent defects using laser thermography. Unfortunately, it is not always possible to clean the surface from contamination, especially for remote objects located in hard-to-reach places.

In addition, the above experimental results do not allow us to unequivocally state that there is a corrosion defect at the interface. At the same time, another approach is possible to solve the problem associated with filtering false test results. In this case it can be assumed that the shape of these thermal radiation pulses is a determining factor in the procedure for detecting and recognizing latent defects.
A computer program for analyzing thermographic images made it possible to correct external contaminants, taking into account the analysis technique for cooling the irradiated surface described in the next section.

4. Computer simulation of thermal processes

The simulation of the results of laser irradiation was carried out on samples of a simplified composition: with a massive base of heat-conducting material (metal) and a surface layer with low heat conductivity (polymer) of various thicknesses, which imitated a surface defect. It is believed that all objects are covered with a thin film of the same protective paint to ensure equal absorption coefficients of laser radiation and emissivity in the spectral range of the recorded thermal signal. As model objects, samples from steel bars with a thickness of H = 60 mm with plastic layers of various thicknesses h were considered.

Laser heating is modeled by the one-dimensional equation of heat conduction along the x axis written for two contacting materials with thicknesses h, H and thermal diffusivity coefficients

IOP Conf. Series: Materials Science and Engineering **904** (2020) 012001 doi:10.1088/1757-899X/904/1/012001

$\alpha_1 = 1.5 \times 10^{-7}$ m^2/s and $\alpha_2 = 1.3 \times 10^{-5}$ m^2/s, respectively.

At the initial time, the temperature of the sample is uniform and equal to T_0. The condition at the interface: $T_1(h) = T_2(h)$, on the surface: $T(h + H) = T_0$.

The problem described by the system of equations must be solved:

$$\begin{cases} \dfrac{\partial T}{\partial t} = \alpha_1 \dfrac{\partial^2 T}{\partial x^2}, & 0 < x < h \\ \dfrac{\partial T}{\partial t} = \alpha_2 \dfrac{\partial^2 T}{\partial x^2}, & h < x < h + H \end{cases} \quad (1)$$

Numerical simulation of the problem was carried out using finite-difference analysis. Fig. 4a shows the graphs of the obtained time dependence of the surface temperature T_{norm} of a number of samples on the dimensionless parameter $k = t/\tau$ (τ is the duration of the pulse exposure) in a normalized form.

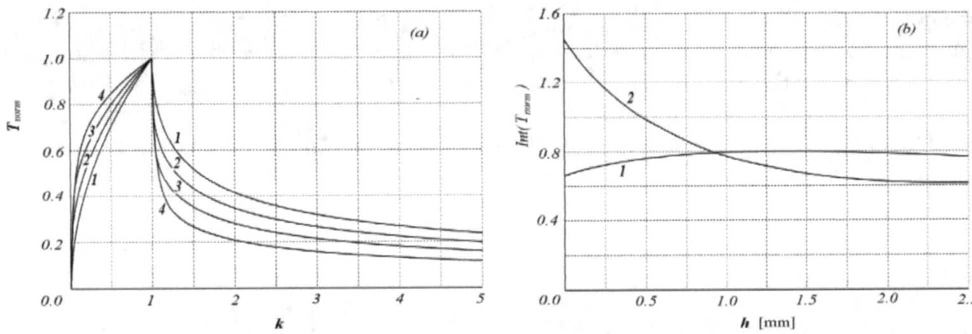

Figure 4a. The calculated time dependences of the relative surface temperatures of the layered structure for different values of h. The data are presented in dimensionless coordinates $k = t/\tau$ for a laser pulse duration of $\tau = 20$ sec. 1– $h = 0$; 2– $h = 0.2$ mm; 3– $h = 0.5$ mm; 4– $h = 1{,}1$ mm.
Figure 4b. Dependences of the integrals of normalized temperatures on the thickness h of the surface layer of polymer: 1 - integral from $k = 0$ to $k = 1$ (heating during a laser pulse), 2 - integral from $k = 1$ to $k = 5$ (cooling)

Figure 4b shows the dependences of the normalized temperature integrals on the thickness h of the surface layer for the heating process (curve 1) and cooling (curve 2). It can be seen that the dependence of the quantity under consideration on h is weak and ambiguous in the region of action of the laser pulse. But when the object cools down, analysis of the integral of the thermal signal can give an unambiguous answer about the value of h.

An example showing the advantage of the described approach is illustrated by the images shown in Fig. 5.

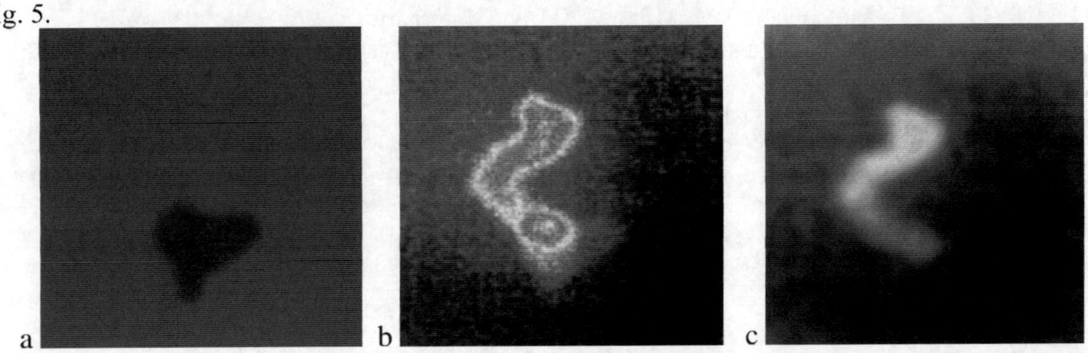

Figure 5. a - surface image in visible light (there is a dark spot on the paint); b – spatial distribution of the thermal signal recorded in pulsed laser thermography of the same sample; c - the result of the analysis of subsurface defects using the described technique

The general view of the sample in visible light is reflected in Fig. 5a. On the surface of the paint coating there is a stain of contamination, which leads to an increase in local absorption of laser radiation. The corresponding zone of increased heating is manifested in the results of pulsed thermography, based only on a comparative analysis of the amplitude characteristics of the heat signal from various parts of the surface of the investigated object (Fig. 5b). This distorts the expected results of a study of subsurface material defects. At the same time, the use of computer processing of a series of consecutive IR images of a sample after exposure to laser heating in accordance with the method of mathematical modeling described above allows identifying precisely surface defects hidden by polymer putty under a common paint layer (Fig. 5c).

With appropriate calibration it is possible not only to detect but also to identify the nature of latent defects in the surface layer of the material.

5. Summary

A new approach to visualization of a hidden defect has been developed by "extracting" it from a digital image of a distorted thermal field caused by heating of surface contamination. This approach is carried out using pulsed laser thermography by analyzing and processing the attenuation curve of the photothermal signal. It is shown that it is possible to detect corrosion products of hydrated iron oxide with a thickness of 100 microns or less at a distance of 10 m.

An experimental technique for detecting hidden surface defects in materials can satisfy the needs of remote monitoring of metal structures in the form of bridges, ships, tanks for transporting and storing petroleum products, etc.

References

[1] P.E. Nordal and S.O. Kanstad: Physica Scripta Vol. 20 (1979), p. 659
[2] G.M. Zverev and L.A. Skvortsov: Izv. Akad. Nauk, USSR, Ser. Fiz. Vol. 45 (1981), p. 644
[3] P.E. Nordal and S.O. Kanstad: Infrared Physics Vol. 25 (1985), p. 295
[4] V.M. Kirillov and L.A. Skvortsov: Quantum Electronics 36 (2006), p. 797
[5] W.J. Parker, R.J. Jenkins, C.P. Butler and G.L. Abbott: Journal of Applied Physics Vol. 32 (1961), p. 1679
[6] S.M. Shepard, in: Thermosense XXIII, Int. Soc. for Opt. and Photonics Vol. 4360 (2001), p. 511
[7] V.N. Lopatkin, O.E. Sidoryuk, L.A. Skvortsov: Sov J Quantum Electron Vol. 15 (1985), p. 216
[8] O. Breitenstein and M. Langenkamp, in: Basics and Use for Functional Diagnostics of Electronics Components (Springer-Verlag Berlin Heidelberg 2003)
[9] X.P. Maldague: Materials Evaluation Vol. 60 (2002), p.1060
[10] J.M. Milne and W.N. Reynolds, in: 1984 Cambridge Symposium, Int. Soc. for Opt. and Photonics (1985), p. 119
[11] S.K. Lau, D.P. Almond and J.M. Milne: NDT & E International Vol. 24 (1991), p. 195
[12] S.M. Shepard, J.R. Lhota and T. Ahmed: Nondestructive Testing and Evaluation Vol. 22 (2007), p. 113
[13] M. Hain, J. Bartl and V. Jacko, in Proc. of the 7th Int. Conf. on Measurement, Bratislava, Inst. of Measurement Science SAS (2009), p. 339
[14] X. Maldague, F. Galmiche and A. Ziadi: Infrared physics & technology Vol. 43 (2002), p. 175

IOP Conf. Series: Materials Science and Engineering **904** (2020) 012002 doi:10.1088/1757-899X/904/1/012002

Development of the solar roof tile with internal cooling

Karel Raz, Zdenek Chval

University of West Bohemia, Regional Technological Institute, Univerzitni 8, 301 00, Pilsen, Czech Republic

kraz@rti.zcu.cz, zdchval@rti.zcu.cz

Abstract. This paper deals with the development of the roof tile with the integrated solar panel. This panel is cooled down with respect to maximizing of the efficiency. Dimensions of the tile are according generally used roof tile. Plastic solar tile can replace the regular one. The design of the tile was validated by usage of thermal simulations and flow simulations in NX Nastran software. These FEM solutions were validated by the measurement on the prototype. The output of this research is prototype from plastic material. This material was selected because of the fast and cheap production.

1. Introduction

Main aim of this paper is design of the roof tile, which is equipped by the cooled solar panel. This panel will be placed on the top side of the tile. Efficiency of the energy conversion from solar to electrical is highly temperature-dependent [1]. Lower temperature leads to the higher electrical output. Designed solar tile has to fully replace regularly used roof tile Tegalit Star [2]. Used cooling medium can be used as source of heat (heating, warming up the water, etc.) in the heat exchanger. The efficiency of solar panels is highly temperature-dependent and it is generally about 13-25%. This efficiency is measured for temperature 25°C and solar loading 1000W/m^2 (according STC standard). The decreasing of output with respect to the temperature is obvious from the following figure (Fig.1). The output (in terms of voltage) around twice lower when temperature of panel is by 100°C higher.

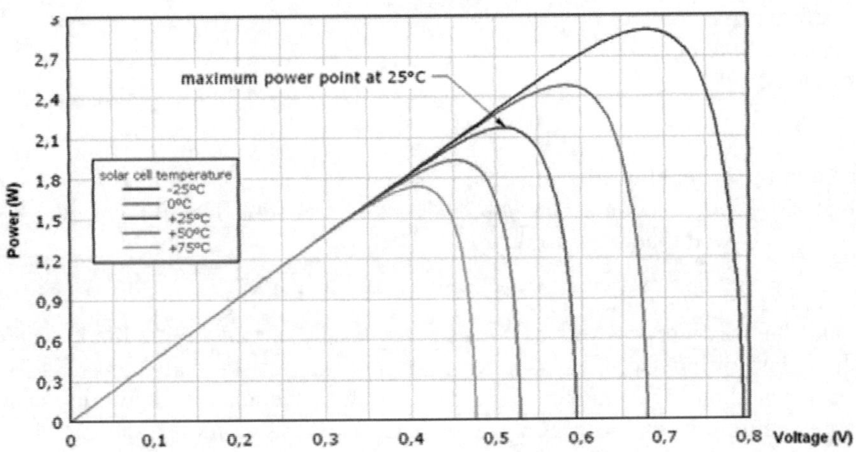

Figure 1. Dependency between power, voltage and temperature of solar panel [3]

Different producers are focused on this area nowadays. Most often is the cooling procedure performed by the water spray focused on the top surface of the solar panel [4,5]. Disadvantage of this solution is in heat losses of cooling water, which is not used further. [6,7,8]

2. Design of the solar roof tile

The design is obvious from the following figure (Fig.2). The tile was created as a plastic part produced by injection molding. The tile has integrated three straight longitudinal cooling channels. Straight design of channels is selected because of minimization of pressure loss. On the top surface of tile is solar panel. This panel is connected by gluing with thermal material as a glue. The flexible hose from PVC performs connection between panels [9,10].

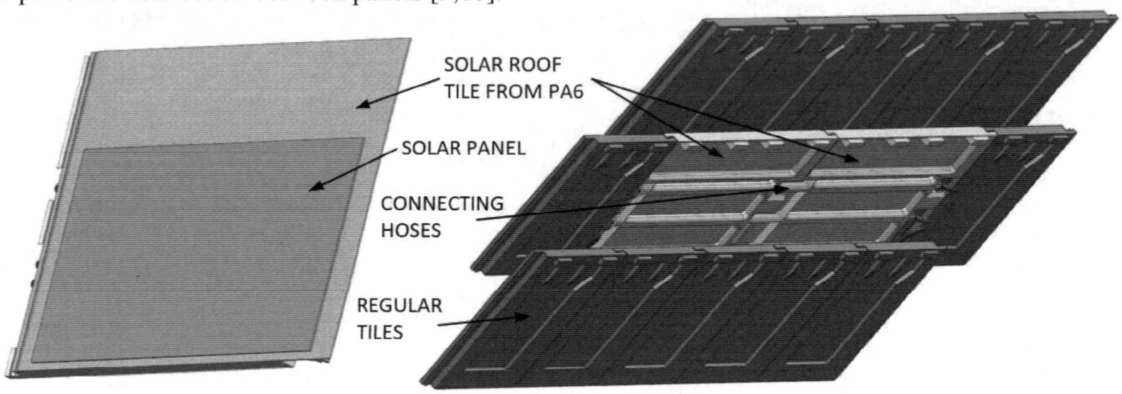

Figure 2. Individual solar tile (left) and array of tiles (right)

3. Pressure loss

The pressure loss is necessary to determine with respect to correct selection of the hydraulic pump and optimal circulation of the cooling liquid. The suitable pump will be used in the experimental device.

3.1. Analytical determination of the pressure loss

Pressure loss of the straight tube is defined as (1): [11,12]

$$\Delta p = \lambda \cdot \frac{l}{d} \cdot \frac{w^2}{2} \cdot \rho \tag{1}$$

Where:
λ – Friction coefficient
w – Average speed of the flow
ρ – Density of the liquid
d – Diameter of the tube

Cylindrical area of the channel is D14x 454 mm. The pressure loss is 0.005 bar.
Hose coupling (two- input and output) has dimensions D8 x 30 mm. The pressure loss is 0.0065 bar.
Analytical solution showed total pressure loss 0.018 bar.

3.2. Virtual simulation of the pressure loss

Determination of the pressure loss was performed also by the virtual simulation by usage of the finite element method. The difference of pressure at input and output was validated (Fig.3). The pressure loss is 0.05 bar. It is about twice higher value comparing to the analytical solution. The difference is caused by more accurate virtual simulation, which is respecting complex shape [13] of cooling and influence of internal friction [14,15].

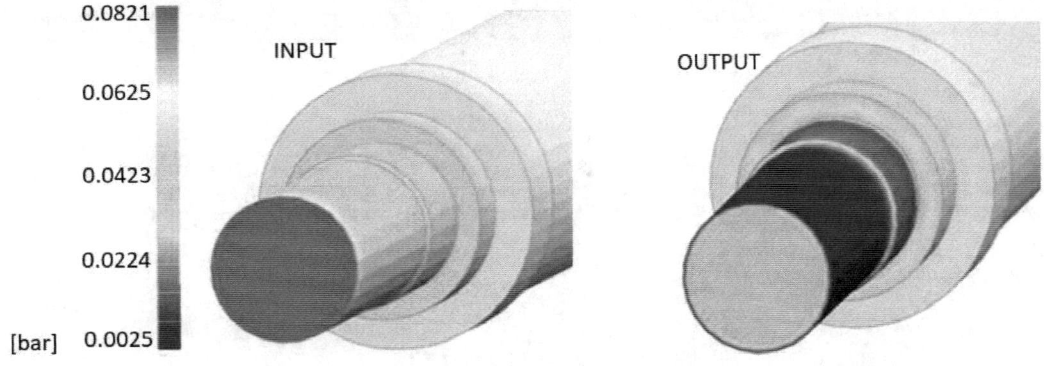

Figure 3. Pressure distribution in fluid at input and output [bar]

4. Validation of the cooling effect

The efficiency of the solar panel was determined for average input values. These values are changed with respect to the period of the year. The simulation was evaluating the average temperature on to top surface of the solar panel. The ambient temperature was selected according the period of year. The loading was constant 800W/m^2. This load was applied on the surface of solar panel and surface of the tile. The influence of clouds and non-uniform sunlight is neglected. The input temperature of cooling water is 20°C, the volume flow is 0.5m^3/h. The material of the tile is PA6 with coefficient of the thermal conductance 0.3 [W·m^{-1}·K^{-1}] [16]. It is obvious that usage of materials with higher thermal conductance is beneficial, but economical reason is more significant [17]. Simulation was performed as transient and steady- state analysis with considering of the dynamics of the fluid. During the simulation was considered the emissivity to the environment. The software NX Nastran was used for the simulation.

4.1. Temperature distribution with cooling- steady state

Average ambient temperature is 16°C. Thermal loading is 800W/m^2 (maximal possible value in Czech Republic).

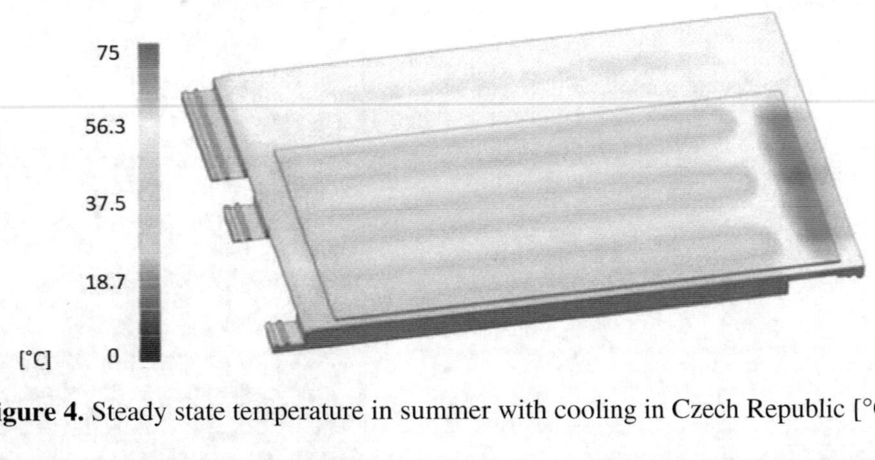

Figure 4. Steady state temperature in summer with cooling in Czech Republic [°C]

Average temperature of the solar panel is 43.1°C (Fig.4). The efficiency of solar panels is 0.4% lower by each degree Celsius. The efficiency is therefore 92.7% comparing to the standardised STC test (25°C and 1000 W/m^2).

4.2. Temperature distribution without cooling- steady state

Average ambient temperature is 16°C. Thermal loading is 800W/m^2 (maximal possible value in Czech Republic).

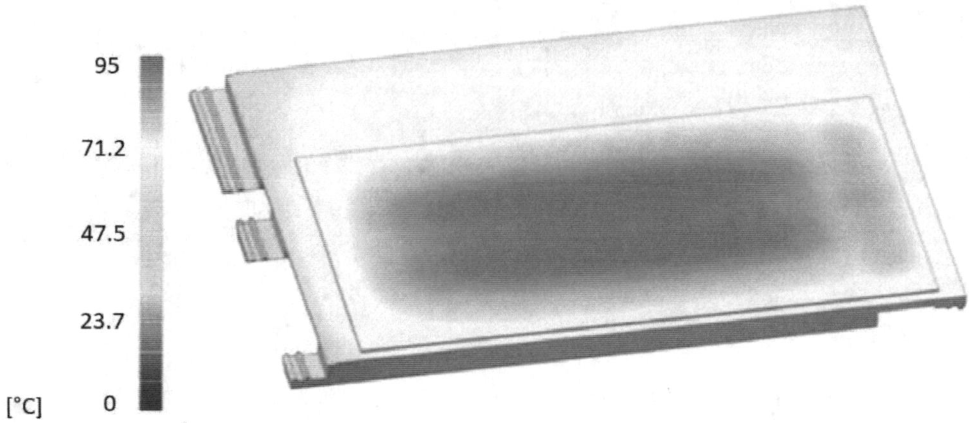

Figure 5. Steady state temperature in summer without cooling in Czech Republic [°C]

The average temperature of the solar panel is 83.2 °C (Fig.5). The efficiency is 76.7% comparing to standardized STC test. The difference between design with cooling and without cooling is about 15%.

5. Measurement of the prototype

Parameters of the testing device are as follows:

Two independent circuit are used, each with 4 panels. One circuit is cooled down by internal circulation of water. Material of roof tiles is PA6. Temperature of each panel and cooling water is measured. The water pump is connected to the battery, so therefore is the prototype independent on the electricity. Electrical load are two bulbs 50W. Angle of device is 30°C (optimal value for Czech Republic). Orientation is to the southwest. The water tank was 40 liters- size. The following figure (Fig.6) shows real prototype and the diagram of connection.

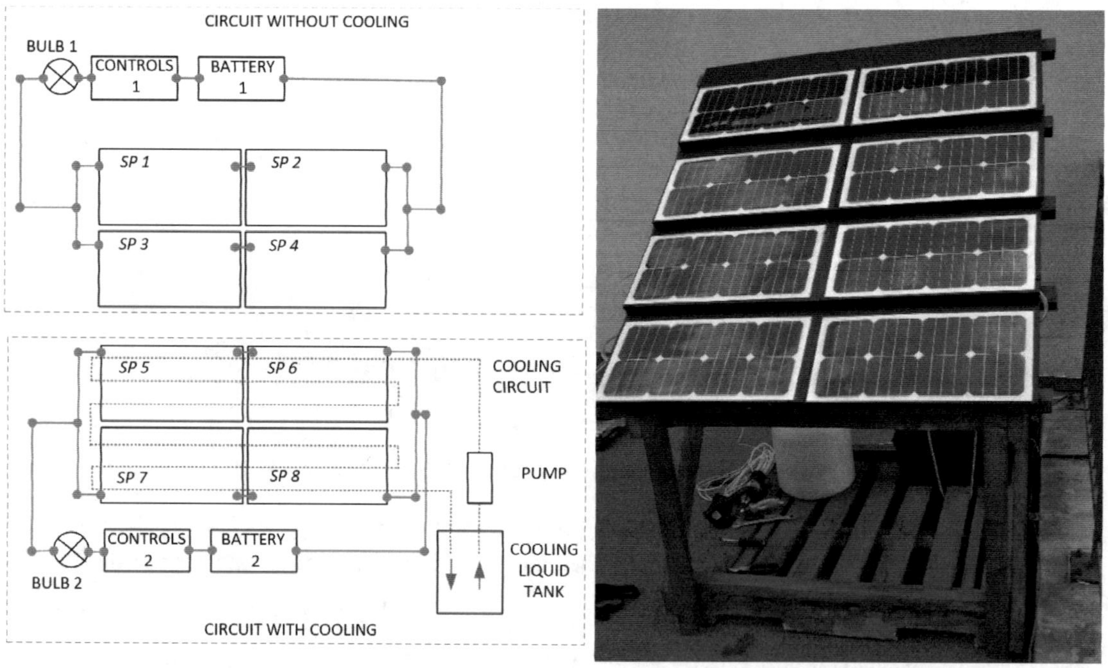

Figure 6. Diagram of the prototype connection (left) and real prototype (right)

The temperature on the upper surface of the solar panel is obvious from the following graph (Fig.7). Thermal cells are located on the plastic material of the roof tile. It is obvious, that is necessary to use the material with better thermal properties (it means with higher thermal conductivity). The difference between both circuits is not so significant. Difference between the virtual simulation and the measurement is caused by lower sunshine and therefore warming up of the panel was much lower during testing.

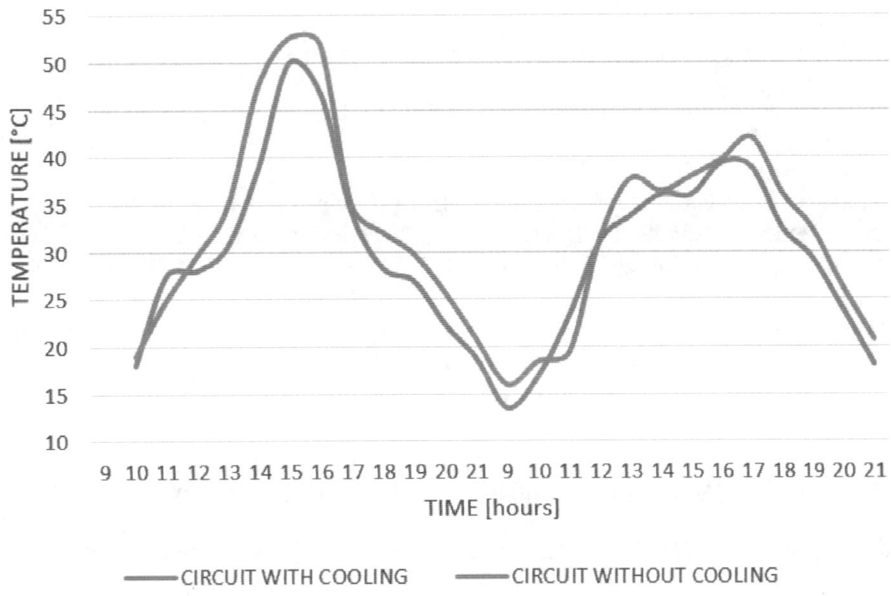

Figure 7. Temperature on the top surface of panels with and without cooling during time period of 11:00-21:00 (in two days)

6. Conclusion

This paper deals with the design of cooled solar tile. The designed solar tile is changeable with the regular tile Tegalit Star. It is possible to produce this solar tile with technology of injection molding. This is suitable because of the economical and time-depended point of view. The roof tile has solar panel on the top surface. Inside the tile are integrated three straight cooling channels. They are improving efficiency of energy conversion.

Mechanical connection between tiles is done by lockers. Hydraulic connection is done by flexible hose. Whole device was tested and temperature was measured on the top surface of tile. The pressure loss is negligible, because is around 0.05 bar. Velocity of cooling water during testing was around 3 m/s. This area of industry is significant, because solar energy is taking key role nowadays. The efficiency of solar panel (depends on individual type) is 0.4% lower by each degree Celsius. Future research will be focused on improving design with respect to better heat transfer. It will be beneficial to use plastic material with thermal additives. The conductivity is then 10times higher.

7. Acknowledgments

This paper has been prepared under project LO1502 'Development of the Regional Technological Institute' under the auspices of the National Sustainability Programme I of the Ministry of Education of the Czech Republic aimed at supporting research, experimental development and innovation.

References

[1] J. Bozzelli, "Injection Molding: Understanding Pressure Loss in Injection Molding", Plastic Technology, December 2010

[2] https://www.bramac.cz/produkty/betonove-tasky/tegalit-star, reached 13.11.2019
[3] K. Lohse, https://www.quora.com/Will-the-solar-panel-gain-more-electricity-if-the-temperature-is-higher, reached 13.11.2019
[4] J. Bozzelli, "Injection Molding: How to Set Second-Stage (Pack and Hold) Pressure", Plastic Technology, February 2011
[5] M. Miller, "Avoiding and Solving Injection Molding Problems Using Shear Rate Calculations-Part 1", Plastics Today, February 2007
[6] J. Bozzelli, "Injection Molding: Develop Guidelines-Not Strict Procedures-For a Robust Molding Process", Plastics Technology, December 2015
[7] A. Ausperger, "Simulation of Deformation and Compression of Fabric During the Back Injection Moulding Process", Annals of DAAAM for 2011 & Proceedings of the 22nd International DAAAM Symposium, 2011, Austria, Volume 22, No. 1, pp. 1261-1262, ISSN 1726-9679, ISBN 978-3-901509-83-4
[8] M. Stanek, M. Manas, D. Manas, K. Kyas, J. Navratil, "Injection Molding Process and its Optimization", Annals of DAAAM for 2011 & Proceedings of the 22nd International DAAAM Symposium, 2011, pp. 0155-0156, Vienna, ISSN 1726-9679, ISBN 978-3-901509-83-4
[9] W. G. Anderson, et al., Heat Pipe Cooling of Concentrating Photovoltaic (CPV) Systems, Conference, Record of the IEEE Photovoltaic Specialists Conference, 2008.
[10] X. Tang, Z. Quan, Y. Zhao, Experimental Investigation of Solar Panel Cooling by a Novel Micro Heat
[11] M. Moradgholi, S. M. Nowee, I. Abrishamchi, Application of heat pipe in an experimental investigation on a novel photovoltaic/thermal (PV/T) system, Solar Energy 107 (2014) 82–88
[12] H. G. Teo, P. S. Lee, M. N. A., Hawlader, An active cooling system for photovoltaic modules, Applied Energy 90 (2012) 309–315
[13] Z. Farhana, et al., Experimental Investigation of Photovoltaic Modules Cooling System, IEEE Symposium on Computers & Informatics 2012.
[14] N. Arcuri, F. Reda, M. De Simone, Energy and thermo-fluid-dynamics evaluations of photovoltaic panels cooled by water and air, Solar Energy 105 (2014) 147–156
[15] Z. Chval, K. Raz, Effect of heat load on a mechanical forging press. In Proceedings of the 27th DAAAM International Symposium, 2016, Vienna: DAAAM International, pp. 344-348. ISBN: 978-3-902734-08-2, ISSN: 1726-9679
[16] V. Marek, Basic Research of Thermal Transfer Simulations, Proceedings of the 27th DAAAM International Symposium, pp.0578-0585, B. Katalinic (Ed.), Published by DAAAM International, 2016, ISBN 978-3-902734-08-2, ISSN 1726-9679, Vienna, Austria, DOI: 10.2507/27th.daaam.proceedings.085
[17] V. Marek, Z. Hajicek, Thermal Simulations Based on Macro-Models, Proceedings of the 28th DAAAM International Symposium, pp.0627-0634, B. Katalinic (Ed.), Published by DAAAM International, 2017, ISBN 978-3-902734-11-2, ISSN 1726-9679, Vienna, Austria, DOI: 10.2507/28th.daaam.proceedings.089

IOP Conf. Series: Materials Science and Engineering **904** (2020) 012003 doi:10.1088/1757-899X/904/1/012003

New decisions in the organization of the procedure for setting norms in standards on the example of metal products

Marina Polyakova[1], Eduard Dryagun[2]

[1] Nosov Magnitogorsk State Technical University, 38, Lenin Avenue, Magnitogorsk, 455000, Russia
[2] OJSC Magnitogorsk hardware and sizing plant "MMK-METIZ" 5, Metiznikov street, Magnitogorsk, 455002, Russia

Corresponding author: m.polyakova@magtu.ru

Abstract. Standard contains the full set of requirements to the product, its production, control and acceptance. A customer and a manufacturer assess one and the same object of standardization in different ways. The function-target analysis makes it possible to find out the relation between the customer functions and the object properties. Application of this developed method results in a network graph reflecting the connections between the main useful function and individual functions as well as the properties providing these functions. The developed approach of assessment of similarity of the customer and the manufacturer positions in the process of development of standard technical requirements is based on the mathematical description of an S-curve. The proposed method makes it possible to take into account the peculiarities for setting norms in standards of metal products.

1. Introduction

One of the most important elements of innovative development is standardization [1 - 3]. It involves both development and setting of requirements to various products; the objects of standardization are operations, processes, services and methods of control and analysis. Standardization can be defined as setting and application of rules in order to normalize activity in certain fields for the mutual benefit and with the participation of all the parties concerned, in particular, to achieve overall optimal saving at the same time meeting all performance conditions and safety requirements. In spite of the obvious achievements in the field of standardization, the process of development and acceptance of specifications of standards is quite controversial.

One of the most important activities in the technological advancement is the development of standards for industrial products. Standardization fulfills different functions at different stages of product life cycle. At the stage of product development, standardization sets the requirements to the quality of the final product on the basis of complex standardization of qualitative characteristics of this product as well as raw materials, semi-finished products and components; besides, standardization makes it possible to develop a unified system of product quality indices depending on the product designation and operating conditions. At the production stage, standards must take into account the nature and characteristics of the final product; they specify the requirements to the control tools and methods and requirements to assessment of the product quality. At the stage of marketing and sales of the product, standardization sets the requirements to package, preservation, transportation and storage.

At the stage of consumption and operation, standardization specifies the unified requirements to maintenance and repair of the product as well as the standards for its utilization.

Analysis shows that the current practice specifies only organizational and legal aspects of the problem of standard development [4 – 6]. However, the development of the list of requirements covered by the standard is an important stage because the level of the standard itself is formed at that very stage.

A standard is the result of teamwork. The standard is developed on the basis of the decision made by the developer (manufacturer) or upon request of the customer of the product and it contains the full set of requirements to the product, its production, control and acceptance. In mass production conditions, the standard becomes a precondition of the exchange of goods. That is why it must express the interests of all the parties of the exchange. In other words, the standard is formed on the basis of the balance of interests of the parties. In practice, this balance is achieved in the negotiation process and in the process of comparison of capabilities and needs of the parties. The main aim here is to ensure the mutual understanding between developers, manufacturers, marketing specialists and customers of products and services. Standards can be a means to achieve this understanding.

However, a customer and a manufacturer assess one and the same object of standardization in different ways. The customer is interested in meeting his needs, that is, in his opinion the product must fulfill certain functions. In practice, the customer makes a choice of the product from the product range, which is on offer in the market. However, he or she is motivated by the need to provide the fulfillment of specific functions. The manufacturer has to make sure that the product possesses certain characteristics in the production process. He considers the product as a set of measurable parameters specified by the standard. The manufacturer has certain equipment, production facilities and production practices. All these factors, to a greater or lesser degree, will limit his capabilities. That is the nature of the conflict of the parties taking part in the development of the product specifications [7 – 9].

At present, the requirements imposed on the object of standardization do not take into account all the needs of the interested parties, and as a result, one of the main principles of standardization is not met, that is, some obstacles to the production process and product handling arise. To solve this problem, it is necessary to formalize the procedure of coordination of views of the interested parties in the process of standard development, which will make it possible to give quantitative assessment of the gap between the positions of the customer and the manufacturer.

2. Methodology

The interaction of the customer and the manufacturer in the process of coordination of the standard requirements to the product can take place in two ways: before the mass production of the product and after the start of the mass production (Figure 1).

The first way of interaction between the customer and the manufacturer means that the coordination of the requirements of the parties takes place before the mass production of the new product; later, these requirements are formalized in the standard. The characteristic feature of this way of interaction between the parties is that the customer and the manufacturer state and coordinate the requirements of the standard to the conceptually new product, which has never been manufactured before. The second way of interaction between the customer and the manufacturer involves adjustment of requirements and characteristics of the product, when they are specified in the standard.

As a rule, the requirements specified in standards are the result of a number of negotiations between customers and manufacturers leading to agreement about the requirements of the standard, which is a trade-off between the parties. A particular standard is negotiated as the optimal combination of customer requirements and manufacturer capabilities. In the present paper, the manufacturer requirements are defined as the requirements and capabilities expressed in quantitative terms and characterized by unit parameters of the product. Customer requirements are defined here as the requirements made by the customer to the product; in the opinion of the customer, these requirements must ensure the product quality corresponding to the functional purpose of the product. That is why it is necessary to choose the method, which will make it possible to establish the links between the product functions and the properties, which provide the fulfillment of these functions [10 – 15]. The analysis shows that most of them are based on the combination of the well-known methods which are used in quality management.

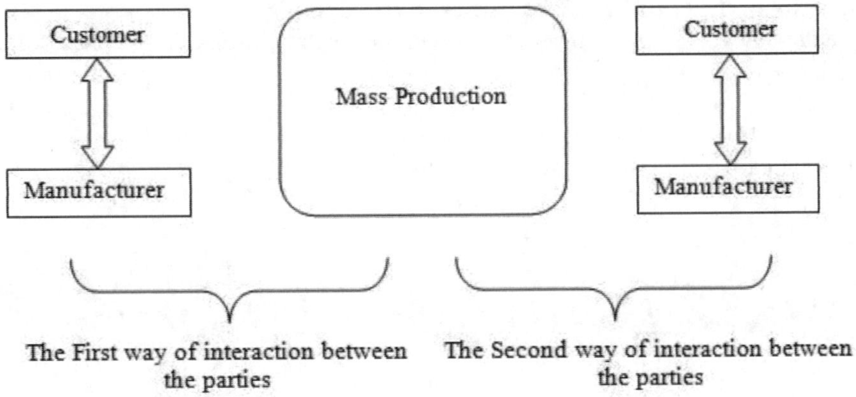

Figure 1. Ways of interaction between the customer and the manufacturer in the process of standard development.

The function-target analysis can be used as the such method [16, 17]. The characteristic property of this method of analysis is that the analysis of the object is carried out from the moment when it acquires the corresponding customer functions and through the whole customer useful phase of the product life cycle without focusing on the method or characteristics of production of this object. The customer useful phase is defined here as the period of time from the date of manufacture of the product to the end of its operation. In other words, this is the period of time when the product is capable of fulfilling its customer functions. Then the process of coordination between the two parties can be divided into two stages: determining the structure of links between the customer functions and the product quality (structural coordination), and the choice of the level of quantitative characteristics providing the coordinated position of the parties (parameter coordination). In this case, parameter coordination can be achieved by maximizing the complex assessment of product properties as a contraction function of property evaluation. Figure 2 shows a scheme of the structural link between the customer functions and product properties.

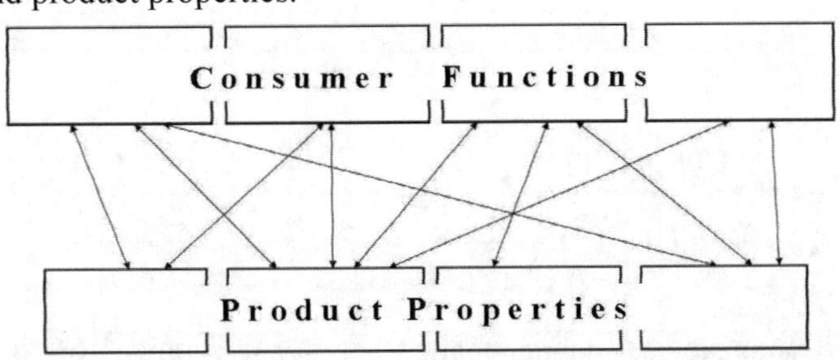

Figure 2. Structural relationship between the product properties and functions.

It is obvious that the customer and the manufacturer specify their requirements in the form of some range of values. Let us consider now the possible variants of overlapping of the ranges of values specified by the customer and provided by the manufacturer. We introduce the following conventional symbols:

F is the range of requirements or capabilities of the manufacturer;

U is the range of customer requirements.

The following principles were used to develop the mathematical model of the degree of similarity between the positions of the parties:

1. If the requirement of the customer *U* cover totally the capabilities of the manufacturer *F*, then the degree of similarity of the positions of the parties *M* is equal to 1.
2. If the ranges of values of requirements specified by the parties do not overlap at all, then the degree of similarity is equal to 0.
3. If the range of requirements of the customer and the manufacturer have some common points the degree of similarity *M* is located in the interval between zero and one $0 \le M \le 1$.

One can mathematically express it in the following way

If $F \subset U$, then $M = 1$. (1)

If $F \cap U = \varnothing$, then $M = 0$. (2)

If $F \cap U \neq \varnothing$ and $F \not\subset U$, then $M = \dfrac{V(F \cap U)}{V(F)}$, $0 < M < 1$. (3)

where M is assessment of the degree of similarity between the positions of the customer and the manufacturer, $V(X)$ is a certain quantitative measure of the set *X*, that is the length of the interval of values of the requirements.

Proceeding from this condition, let us define the assessment function of the interval measure $M(V(F \cap U))$ in the general case. It will define the following calculation algorithm for the rate of assessment change as a function of the parameter value

If $V(F \cap U) = 0$, then $M(0) = 0$. (4)

If $V(F \cap U) = V(F)$, then $M(V(F)) = 1$. (5)

If $\dfrac{V(F)}{2}$, then $M\left(\dfrac{V(F)}{2}\right) = \dfrac{1}{2}$. (6)

If $0 \le V(F \cap U) \le \dfrac{V(F)}{2}$, then $M(V(F \cap U)) = \dfrac{2 \cdot V(F \cap U)^2}{V(F)^2}$. (7)

If $\dfrac{V(F)}{2} \le V(F \cap U) \le V(F)$,

then $M(V(F \cap U)) = 1 - \dfrac{(V(F) - V(F \cap U))^2}{V(F) \cdot (V(F) - \dfrac{V(F)}{2})} = 1 - \dfrac{2 \cdot (V(F) - V(F \cap U))^2}{V(F)^2}$. (8)

By substituting different values of discrepancy between the ranges of the customer *U* and the manufacturer *F*, we can obtain the quantitative assessment of the degree of similarity of the positions of the parties $M(V(F \cap U))$ for arbitrary values of $V(F \cap U)$.

3. Interpretation, results and discussion

Let us consider the application of the developed approach using high-strength reinforcement for reinforced-concrete structures as an example. Reinforcement is an important component of reinforced concrete, that is why it has to meet the following requirements: it must show good performance characteristics in combination with concrete at all stages of the structure useful life; it must be used to the yield point in case when the bearing capacity of the structure is exhausted; it must provide operating convenience of reinforcement installation and provide the possibility of its mechanization. Then, in accordance with the function-target analysis, the relationship between the useful functions and the properties of high-strength reinforcement at various stages of its useful life can be represented in the form of a network structure (Figure 3).

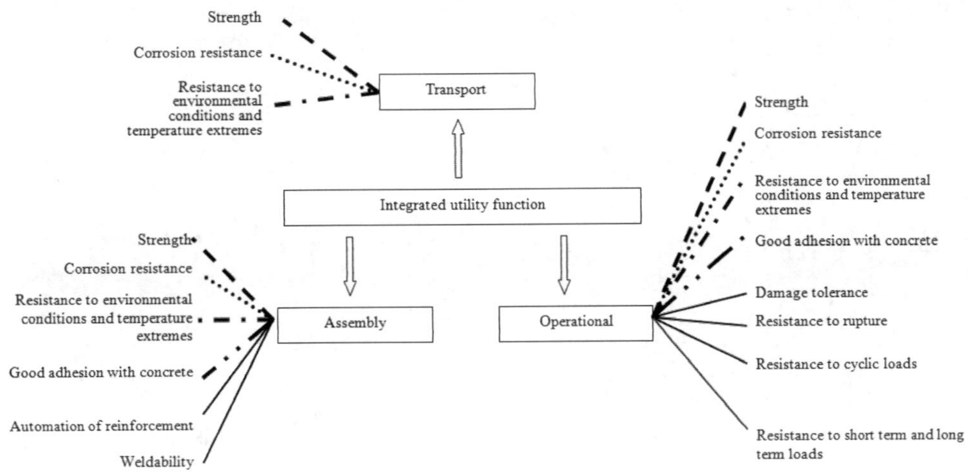

Figure 3. Network structure of relationship between customer functions and properties of high-tensile reinforcement for reinforced-concrete structures.

However, one should take into account that in the process of negotiating the requirements of the standard for the product, the customer and the manufacturer can express their initial requirements in different ways. In the first case, the customer expresses the requirements clearly, that is, he is fully aware of the functions which must be fulfilled by the product. In this case, one can establish the links between the customer functions and the properties at once. However, in some cases, the customer can form the requirements, which are not specific. In other words, he specifies only the necessary property, and it is not clear how the manufacturer can provide this property. For example, the high strength of reinforcement can achieved by making use of certain chemical composition, by special processing modes or by nanostructuring of the semi-finished product. One more variant is possible, when the customer specifies the properties of the product, which further determine the customer properties.
On the basis of the developed principles of mathematical assessment of the similarity of positions of the interested parties and taking into account the continuity of the process, one can represent it in the form of an S-curve (Figure 4) diagramme.

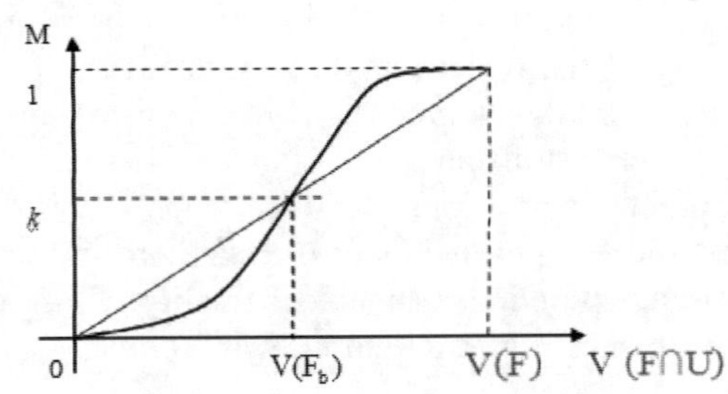

Figure 4. S-curve diagramme estimating the similarity of positions of the parties in the process of setting norms for standard requirements.

This curve can be described mathematically, because it has two characteristic sections: the sections of accellerated and slow growth [18-20]. The bending point is the moment when the rate of growth is equal to the rate of slowing down of the process development. Then this curve can be represented in the form of the following set of formulas

$$
\begin{cases}
0, \text{ when } V(F\,\mathrm{I}\;U)=0. & (9)\\
0.5, \text{ when } V(F\,\mathrm{I}\;U)=V(F_b). & (10)\\
M(V(F\,\mathrm{I}\;U))=1, \text{ when } V(F\,\mathrm{I}\;U)=V(F). & (11)\\
\dfrac{(V(F\,\mathrm{I}\;U))^2}{V(F)\cdot V(F_b)}, \text{ when } 0\le V(F\,\mathrm{I}\;U)\le V(F_b). & (12)\\
1-\dfrac{(V(F)-V(F\,\mathrm{I}\;U))^2}{V(F)\cdot(V(F)-V(F_b))}, \text{ when } V(F_b)\le V(F\,\mathrm{I}\;U)\le V(F). & (13)
\end{cases}
$$

The developed approach can be used to form technical requirements of the standard for the product. Then the main stages of this process can be stated as:

1. Selection of the set of unit quality parameters of the product (manufacturer requirements).
2. Defining the customer functions of the product (customer requirements).
3. Matching the customer functions to the unit quality parameters of the product (function-target analysis).
4. Coordination of matching the customer requirements and the manufacturer requirements.
5. Determining the overlapping of ranges of values of product quality parameters specified by the customer and provided by the manufacturer.
6. Plotting an S-curve and calculation of assessment of similarity of the customer and manufacturer positions.
7. Calculation of group assessment of similarity of parties' positions (if necessary).
8. Calculation of integrated assessment of similarity of the parties' positions (if necessary).

4. Summary

1. The developed approach of assessment of similarity of the customer and the manufacturer positions in the process of development of standard technical requirements is based on the mathematical description of an S-curve. The proposed method makes it possible to take into account the peculiarities of regulatory activity of product properties in the form of interval values. Application of mathematical tools makes it possible both to define accurately the positions of the customer and the capabilities of the manufacturer and to find the mutually acceptable way of regulating the unit quality parameters of the product in the process of standard development; the decision-making time is reduced too.
2. The main methodological principles of standardization are the structural and parameter coordination of the customer and the manufacturer positions. This approach allowed us to develop a mathematical model of coordinating the technical requirements of the standard in the form of optimization problem with the target function of the integrated assessment of quality. Close-cut separation of the customer and the manufacturer positions makes it possible to indicate the ways of promotion for new promising products.
3. The developed principles of structure and parameter coordination are the basis of all the common methods and principles of standardization, such as unification, balance of interests, consistency and others. The concept of advanced standardization complies with principles too. At present by advanced standardization we understand standardization, which sets up higher level of standards and

requirements as compared with the one already achieved in practice, and according to forecasts, these higher standards and requirements will be appropriate in the near future. However, this definition implies changes only in parameters. Structural development of standards is another promising way of advanced standardization. In our opinion, the developed approach to the process of standardization offers a new scientific foundation for developing of this process.

5. Acknowledgments

The reported study was funded by RFBR according to the research project №18-58-45008 IND_a..

References

[1] R. H. Allen and R. D. Sriram: Technological Forecasting and Social Change Vol. 64 (2–3) (2000), p. 171 – 181
[2] Yu. Sered and Y. Reich: J Comp-Aided Design Vol. 38 (5) (2006), p. 405-416
[3] K. Blind and N. Thumm: Res Pol Vol. 33 (10) (2004), p. 1583-1598
[4] M.B. Spring: Int J Standard and Res Vol. **14 (1) (2016), p.** 34-44
[5] K. Blind and A. Jungmittag: J Product Analysis Vol. 29 (2008), p. 51-60
[6] G. Dobrescu and Y. Reich: J Comp-Aided Design Vol. 35 (2003), p. 791–806
[7] M.G. Filho and E.V. Saes: Int J Adv Manuf Tech Vol. 64(5) (2013), p. 1177–1191
[8] C-C. Chen and W-Y. Cheng: Int J Adv Manuf Tech Vol. 34 (2007), p. 1236-1245
[9] W. Dai, P.G. Maropoulos and X.Q. Tang: Proceedings 36th Int MATADOR Conf (2010), p. 145-148
[10] Y. Yao, L. Zhao and Y. Qin: Appl Mech Mater Vol. 37-38 (2010), p. 905-909
[11] Z. Sener and E.E. Karsak: Int J Adv Manuf Tech Vol. 48 (2010), p. 1173-1184
[12] Z. Jiang: Appl Mech Mater Vol. 263-266 (2013), p. 839-842
[13] L. Jun, K. Shulin and L. Pengyu: Key Eng Mater Vol. 467-469 (2011), p. 2103-2108
[14] Z. Yu and J. Zhou: Adv Mater Res Vol. 214 (2011), p. 612-617
[15] R. Schmitt, S. Stiller and B. Falk: Enabling Manufacturing Competitiveness and Economic Sustainability (2014), p. 309-314
[16] M.A. Polyakova, G.Sh. Rubin, G.S. Gun and Yu.V. Danilova: CIS Iron and Steel Review Vol. 12 (2016), p. 45-48.
[17] M. Polyakova and G. Rubin: AIP Conf Proc Vol. 1858 (2017), At. no 040005.
[18] G. Rubin, M. Polyakova and G. Gun: Proc 2015 Int. Conf. on Modeling, Simulation and Applied Mathematics Vol. 122 (2015), p. 178 - 181
[19] M. Polyakova, G. Rubin, G. Gun and Yu. Danilova: Int J Quality Res Vol. 12(3) (2018), p. 573-592
[20] G. Rubin, M. Polyakova, G. Gun: Vestnik of Nosov Magnitogorsk State Technical University Vol. 1 (2015), p. 70-75

IOP Conf. Series: Materials Science and Engineering **904** (2020) 012004 doi:10.1088/1757-899X/904/1/012004

Stress distribution in highly porous SiO2 films: results of the molecular dynamics simulation

F.V. Grigoriev[1,2,a], V.B. Sulimov[1,2,b] and A.V. Tikhonravov[1,2,c]

[1]RCC M.V. Lomonosov MSU, Russian Federation
[2]Moscow Center of Fundamental and Applied Mathematics, 119234, Leninskie Gory, 1, Moscow, Russia

[a]fedor.grigoriev@gmail.com, [b]vs@dimonta.com, [c]tikh@srcc.msu.ru

Abstract. The stress distribution over the thickness of silicon dioxide thin films is studied using simulated thin film cluster. The atomistic cluster representing the film is deposited on a glassy substrate using the full-atomistic molecular dynamic simulation. The deposition angle is equal to 80°, which leads to the growth of a highly porous anisotropic film with low refractive index. The method for calculating the stress distribution is based on the integral relationship between the thickness-averaged stress and the stress distribution. In the present work, we focus on the application of this relationship to the atomistic modeling of stresses at the initial stage of thin film growth. It is found that in the transition layer between the substrate and the film the stress distribution function corresponds to the compressive stress. With the increase in film thickness, the stress distribution function changes sign and becomes tensile. It is shown that these results correspond to experimental data.

1. Introduction

Mechanical stresses arising in growing films significantly affect the optical coatings quality [1]. These stresses depend on the deposition conditions. One of the key parameters affecting the properties of thin films is the deposition angle α. Normal deposition corresponding to the low values of α leads to the formation of dense and homogeneous structures. Deposition at angles greater than approximately 60° leads to the formation of highly porous films [2]. This deposition regime is called glancing angle deposition (GLAD). Due to the low reflectance and anisotropy of the refractive index, GLAD films are interesting for various applications in optics [3].

Due to the progress in high performance computing, it is now possible to perform a full-atomistic molecular dynamic (MD) simulation of the thin film deposition process. Based on the results of this simulation, various structural and mechanical parameters of the film can be calculated. In the present work, the stresses in the GLAD silicon dioxide films are investigated in the framework of a previously developed MD approach [4-7]. High-performance parallel computing allows to achieve film thicknesses of about 50 nm which is close to the technological thickness of one layer in multilayers optical coatings.

The stress distribution function, or in-plane stress, is defined as the stress in a thin layer parallel to the substrate plane. Layer position is specified by the vertical coordinate h. The thickness-averaged stress for different values of film thicknesses is calculated based on the results of the MD simulation of the deposition process. The dependence of the in-plane stress on vertical coordinate is determined through the derivative of the product of averaged stresses and film thickness with respect to the vertical coordinate [1].

2. Method

The deposition of SiO_2 films is simulated using the step by step MD-based procedure [4-7]. At each injection step, silicon and oxygen atoms with stoichiometric proportion of 1:2 are inserted randomly at the top of the simulation box. Energy of the interatomic interaction is calculated in the frame of the empirical pairwise DESIL force field [6]. The next functional form of this force field is used:

$$U = q_i q_j / r_{ij} + A_{ij}/r_{ij}^{12} - B_{ij}/r_{ij}^{6} \quad (1).$$

Here $q_{i(j)}$ is the charge of $i(j)$-th atom, $q_O = -0.5q_{Si} = -0.65e$, A_{ij} and B_{ij} are parameters of Lennard-Jones potential for the van der Waals interaction, r_{ij} is the interatomic distance, $A_{SiO} = 4.6 \cdot 10^{-8}$ kJ·(nm)^{12}/mol, $A_{SiSi} = A_{OO} = 1.5 \cdot 10^{-6}$ kJ·(nm)^{12}/mol, $B_{SiO} = 4.2 \cdot 10^{-3}$ kJ·(nm)^{6}/mol, $B_{SiSi} = B_{OO} = 5 \cdot 10^{-5}$ kJ·(nm)^{6}/mol.

The total number of silicon and oxygen atoms per injection step is equal to 180. The initial values of the silicon and oxygen atoms velocities correspond to the deposited atoms kinetic energies E(Si) = 10 eV and E(O) = 0.1 eV. These values are typical for the high-energy deposition processes like ion beam sputtering [8].

At each injection step, the *NVT* (constant number of particles, volume and temperature) ensemble with the periodic boundary conditions is used. The vertical dimension of the simulation box increases every injection step by $\Delta h = 0.01$ nm to compensate for the increase in film thickness. The horizontal dimensions of the box remain unchanged. The Berendsen thermostat [9] is applied to keep the simulation cluster temperature constant at $T = 300$ K. The duration of one injection step is 6 ps and the time step of MD modeling is 0.5 fs.

Before the simulation of the deposition process the fused silica substrate is prepared by the melting-quenching procedure from an alpha-quartz crystal [7]. Horizontal dimensions of substrate clusters are equal to 20 nm and 60 nm, the vertical dimension is equal to 6 nm. Horizontal dimensions of the cluster in the simulation of film deposition are the same as those of the substrate. The deposition angle is equal to 80° to obtain GLAD film.

The calculation of the thickness-averaged stresses is organized as follows:

1. Simulation of the silicon dioxide film deposition. The number of injection steps is $N_s = 500$. This ensures a growth of layer with the thickness of approximately 5 nm.
2. Simulation of the deposited cluster in the *NVT* ensemble and calculation of the pressure tensor components $p_{xx(yy)}$. The initial state for the simulation is the final state after the completion of N_s injection steps. The length of the MD trajectory for averaging of $p_{xx(yy)}$ values is taken equal to 100 ps. The thickness-averaged stress tensor components are calculated as follows:

$$\sigma_{xx(yy)} = - p_{xx(yy)} \cdot (L/H) \quad (2),$$

where L and H are the vertical dimension of the simulation box and film thickness, respectively (Fig. 1). The multiplier (L/H) in Eq (2) provides a pressure correction taking into account the empty volume at the top of the box.

3. Stages 1-2 are repeated until the film thickness achieves a specified value. Achieving the thickness of about 40 nm requires approximately 4000 injection steps. Total number of atoms in the final cluster is about two million.

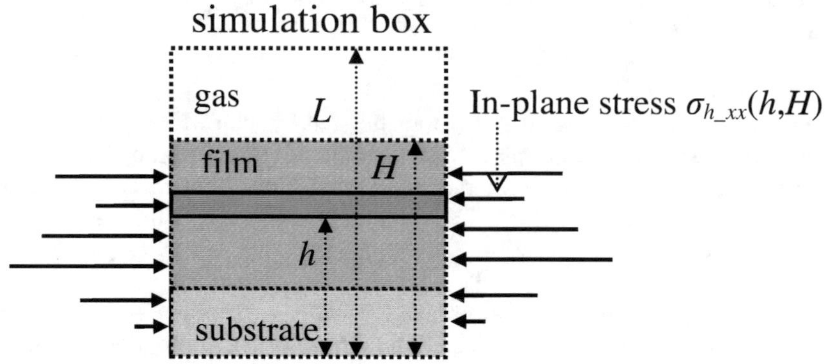

Figure 1. To the calculation of the stress tensor components.

The stress distribution function (in-plane stress) $\sigma_{h\_xx(yy)}(h)$ throughout the vertical coordinate h is related to thickness-averaged stress $\sigma_{xx(yy)}(H)$ by the following equation [1]:

$$\sigma_{xx(yy)}(H) = \frac{1}{H}\int_0^H \sigma_{h\_xx(yy)}(h)dh \tag{3}$$

where H is the thickness of the deposited film, $h < H$. Multiplying both parts of the Eq (3) by H and differentiating with respect to H we obtain the relaitionship between the thickness-averaged stress and the in-plane stress:

$$\frac{d}{dH}(H\sigma_{xx(yy)}(H)) = \sigma_{h\_xx(yy)}(H) \tag{4}$$

MD simulation is performed using the GROMACS package [10]. Components of the pressure tensor are calculated using the kinetic energy and virial tensors [10]. For visual analysis of the atomistic structures the Visual Molecular Dynamic (VMD) package is used [11]. The simulation is carried out using equipment at the shared research facilities of high performance computing resources at Lomonosov Moscow State University [12].

3. Results and discussion

The resulting deposited structure is shown in the central part of Fig. 2.

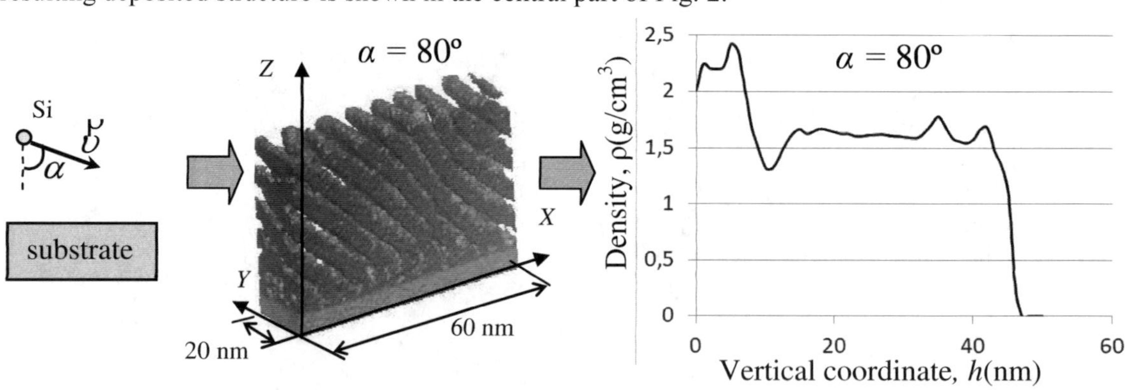

Figure 2. Geometry of the simulation box, atomistic structure of the deposited GLAD film and density profile of the deposited film.

The deposition under the angle $\alpha = 80°$ leads to the formation of a high porous anisotropic structure. The density profile (right side of Fig. 2) shows a significant decrease in the density of GLAD film compared to the density of the normally deposited film, equal to 2.4 g/cm^3. This structure is typical for the deposition under large values of α [1,2].

The results of the average stress calculations are shown in Fig. 3. At the initial stage of film deposition the stress in compressive. With an increase in the film thickness, a shadow effect [1] appears, which leads to the formation of separated slanted columns and voids between these columns in the film structure (Fig. 2, central part). The voids absorb stresses arising in the film due to interaction with deposited high-energy Si atoms. This leads to a decrease in the absolute values of the stress and a change in the sign of stress of σ_{yy} component. The of σ_{xx} remains compressive in the studied interval of film thickness.

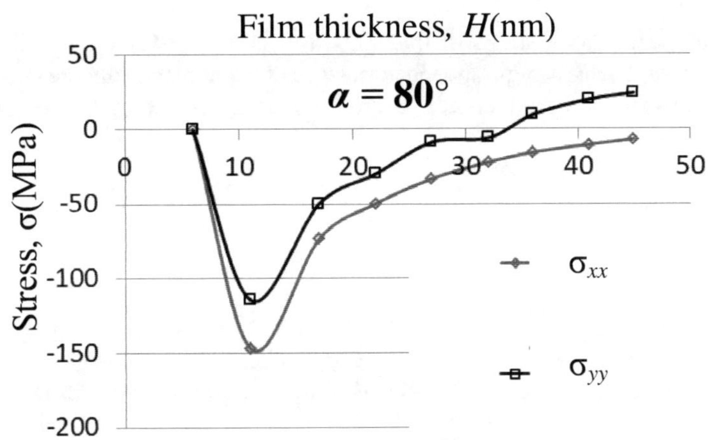

Figure 3. Dependencies of the thickness-averaged stress tensor components σxx(yy) on the film thickness.

The components of the in-plane stress tensor are calculated using Eq (4). The results are shown in Fig. 4.

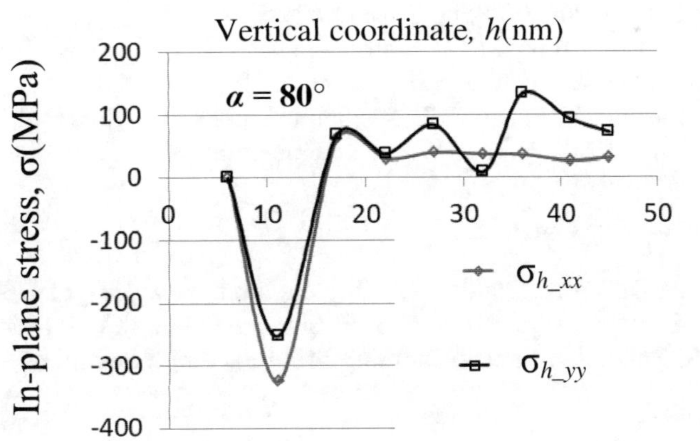

Figure 4. Dependencies of the in-plane stress tensor components on the vertical coordinate h, α is the deposition angle. The total thickness of the deposited film and substrate is 45 nm.

If the vertical coordinate is less than 15 nm, both components of the stress tensor are compressive. The transition layer between the substrate and film makes a significant contribution to the volume-averaged film stress. With further vertical coordinate growth, the stress becomes tensile for both components. The change of the stress sign with growth of vertical coordinate was observed in the experiments [1].

The difference in the values of σ_{h_xx} and σ_{h_xx} indicates the anisotropy of the GLAD film properties. This is due to the anisotropy of the film structure, see the central part of Fig. 3.

4. Conclusion

The distribution of the in-plane stress in the GLAD silicon dioxide film has been studied using thin film clusters simulated by the previously developed MD based method. The method of the in-plane stress calculation is based on the integral relationship between the thickness-averaged stress and the in-plane stress.

It is shown that the in-plane stress in the transition layer between the substrate and the film is compressive. The stress changes sign and becomes tensile when the thickness of the deposited layer exceeds approximately fifteen nanometers. This result is in agreement with the available experimental data.

5. Acknowledgement

The work is carried out according to the research program of Moscow Center of Fundamental and Applied Mathematics. The work was supported by the Russian Science Foundation (grant number 19-11-00053).

References

[1] G. Abadias, E. Chason, J. Keckes, M. Sebastiani, G. B. Thompson, E. Barthel, G. L. Doll, C. E. Murray, C. H. Stoessel, L. Martinu: J. of Vac. Sci & Tech. A Vol. 36 (2018), 020801

[2] M. M. Hawkeye, M. J. Brett: J. Vac. Sci. Technol. Vol 25 (2007), p. 1317.

[3] A. Barranco, A. Borras, A. R. Gonzalez-Elipe, A. Palmero: Progress in Materials Science Vol. 76, (2016), p. 59

[4] F.V. Grigoriev, V.B. Sulimov, A.V. Tikhonravov: J. of N.-Cr. Sol. Vol. 512 (2019), p. 98

[5] F.V. Grigoriev, V.B. Sulimov, A.V. Tikhonravov: Coatings. Vol. 9 (2019), p. 568-576

[6] F.V. Grigoriev, A. V. Sulimov, I. V. Kochikov, O. A. Kondakova, V. B. Sulimov, A. V. Tikhonravov. Supercomputer modeling of the ion beam sputtering process: full-atomistic level, in Proc. of SPIE – the international society for optical engineering, Jena, Germany ,2015. V. 9627. pp. 962708-1-962708-9, doi:10.1117/12.2190938

[7] F. V. Grigoriev, A. V. Sulimov, E. V. Katkova, I.V. Kochikov, O.A. Kondakova, V. B. Sulimov, A. V. Tikhonravov: Applied Optics. Vol. 56 (2017), C87-C90.

[8] A. Tabata, N. Matsuno, Y. Suzuoki, T. Mizutani: Thin Solid Films Vol. 289 (1996), p. 84

[9] H. J. C. Berendsen, J. P. M. Postma, W. F. van Gunsteren, A. DiNola, J. R. Haak: J. of Chem. Phys. Vol. 81 (1984), p. 3684

[10] M. J. Abraham, T. Murtola, R. Schulz, S. Páll, J. C. Smith, B. Hess, E. Lindahl: SoftwareX, Vol. 1-2 (2015), p.19.

[11] W. Humphrey, A. Dalke, K. Schulten: J. Molec. Graphics Vol. 14 (1996), p. 33

[12] Vl. Voevodin, A. Antonov, D. Nikitenko, P. Shvets, S. Sobolev, I. Sidorov, K. Stefanov, Vlad. Voevodin, S. Zhumatiy: Supercomputing Frontiers and Innovations Vol. 6 (2019), p. 4.

IOP Conf. Series: Materials Science and Engineering **904** (2020) 012005 doi:10.1088/1757-899X/904/1/012005

Fabrication of smart supramolecular systems based on polyelectrolytes and surfactants in microfluidic environment

Artem Bezrukov[1], **Yury Galyametdinov**[1]

[1]Kazan National Research Technological University, 68 Karl Marx str., Kazan, Russia

artem.bezrukov.kazan@gmail.com

Abstract. This work represents a process of fabrication of smart materials in a microfluidic confinement. Compared with macroscopic solutions, microfluidic convection-diffusion conditions create multiple additional factors that affect complexation in reacting polymer and surfactant flows. The system of convection-diffusion-reaction equations was solved to develop a polyelectrolyte-surfactant interaction model in a microchannel with laminar flows of reactants; phase formation and precipitation conditions in a microchannel were considered. The recommendations for control of phase behaviour in a microchannel were proposed by the design and operation mode of microfluidic devices used for synthesis of polymer-surfactant complexes and characterization of polymer-surfactant association process.

1. Introduction

Polymers and surfactants are typical representatives of complex supramolecular systems [1-3]. Their association behavior and properties can be controlled by various factors such as concentration ratios, individual characteristics of components, addition of solvent, etc. [4-6].

Microfluidic chips are becoming promising devices for fundamental and applied research for characterization of soft matter such as drug delivery systems, quantum dots, complexes of biopolymers, etc. [7-11]. In microfluidic chips, application of a broad variety of factors can be reduced to control of by microchip's design and operation. There is a growing interest of scientists to synthesis various smart supramolecular systems such as polymer-surfactant associates in microfluidic confinement [12-15].

It is of fundamental and practical importance to reveal operation modes and design parameters of microfluidic devices, which allow for controlling association processes in a microchannel.

The main objective of this research was to study association processes and hydrodynamic properties of resulting complexes for the anionic polyelectrolyte (polyacrylic acid – PAA) and cationic surfactant (cetyltrimethylammonium bromide – CTAB) solutions in microfluidic chips.

2. Experimental

The molds for microfluidic chips were fabricated by deposition of a negative photoresist SU-8 (MicroChem, USA) on a silicon wafer by spin-coating and further treatment by the UV radiation through a photomask according to SU-8 specifications.

The microfluidic devices were fabricated from polydimethylsiloxane (PDMS) Sylgard 184 by Dow Corning, USA. A required amount of PDMS was mixed with a curing agent according to the manufacturer's specifications. The mold was filled with PDMS to form a layer of 3-4 mm. A plastic

Petri dish was then filled with remaining polymer to form a bottom layer with 2-3 mm thickness. PDMS was cured at 60ºC for 4 hours. Cured PDMS layers was removed from a mold and a Petri dish and bonded together. Before bonding, PDMS surfaces were processed surfaces in oxygen plasma in Harrick Plasma Cleaner PDC-32G for 60 seconds.

Association of polymer-surfactant complexes in microchannels were characterized by Levenhuk D320 optical microscope equipped with Levenhuk M1400 Plus camera.

3D-modelling of polyelectrolyte-surfactant interaction in a microchannel was simulated in Matlab 2020a with partial differential equations toolbox.

Diffusion coefficients and hydrodynamic radii of polyelectrolyte macromolecules and polyelectrolyte-surfactant associates were determined by dynamic light scattering with Malvern Zetasizer Nano ZS.

Polyacrylic acid (PAA) by Polysciences, Inc., USA, with Mw = 90 000, 25 w. % aqueous solution was used for an anionic polyelectrolyte. Cetyltrimethylammonium bromide, 98% by BDH Limited, Great Britain (CTAB) was used for a cationic surfactant. Aqueous CTAB solutions with 14 mmol/l surfactant concentration prepared from the CTAB powder were used in all experiments.

3. Theory

A general scheme of a microfluidic chip with reacting polymer, surfactant and solvent flows is shown on Fig. 1. This figure shows a combination of a microchip microscopy photo and the model of convection-diffusion-reaction processes occurring in a microchannel after junction of the inlet flows.

Diffusion of components to the center of the microchannel from the side flows results in the start of the complexation reaction and accumulation of the polymer-surfactant complex in the center of a microchannel.

The widths of polymer, surfactant and solvent flows are the Dirichlet boundary conditions for the junction face of the microchannel parallelepiped. Top, bottom and side faces of the microchannel are the Neumann boundary conditions reflecting zero flux of interacting particles through channel boundaries.

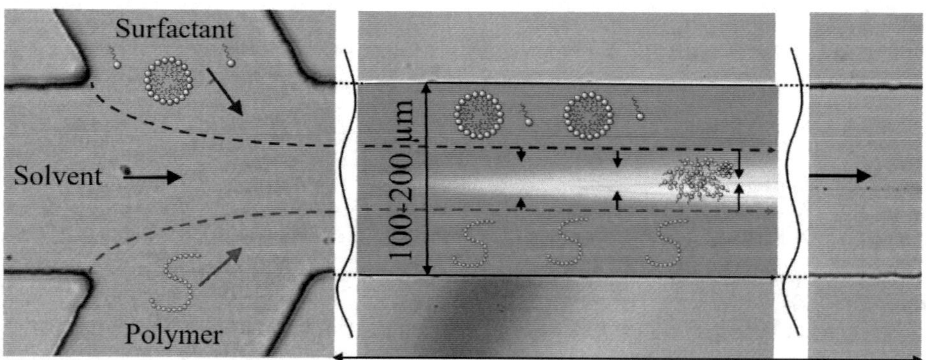

Figure 1. A scheme of a 3-input microfluidic chip with reacting polymer and surfactant flows.

The behavior of reacting flows in a microchannel can be described by a combination of reaction, convection and diffusion partial differential equations with non-uniform initial conditions [16-18]. Association of a surfactant molecule with a monomer binding site in a macromolecular chain can be described by a general second-order reaction equation:

$$A + B \underset{k_2}{\overset{k_1}{\leftrightarrow}} C \tag{1}$$

Where A and B are the reagents A and B, C is the product of their interaction. The rate law for the reagent A is:

$$\frac{dC_A}{d\tau} = -k_1 C_A C_B + k_2 C_C \quad (2)$$

Where C_A, C_B and C_C are concentrations of the reagents A, B and the product C; k_1 and k_2 are the rate constants of the direct and reverse reactions.
For a reaction occurring in a microfluidic channel, the diffusion term and the convection terms are the important components of the equation:

$$\frac{dC_A}{d\tau} + U\frac{\partial C_A}{\partial l} = D_A \Delta C_A - k_1 C_A C_B + k_2 C_C \quad (3)$$

Where D_A is the diffusion coefficient of the reagent A molecules, U is the flow velocity derived from the flowrate, Δ is the Laplacian, and l is the axial coordinate a microchannel.
To simplify this equation, consider a steady-state flow if reactants and two second-order derivatives representing radial and vertical diffusion in a microchannel (the axial diffusion is neglected):

$$U\frac{\partial C_A}{\partial l} = D_A\left(\frac{\partial^2 C_A}{\partial w^2} + \frac{\partial^2 C_A}{\partial h^2}\right) - k_1 C_A C_B + k_2 C_C \quad (4)$$

These equations are non-linear second-order convection-diffusion equations with the chemical reaction added as a source term. Such equations can be derived for the reagent B and the product C. The resulting system of such equations will characterize the behavior of polymer-surfactant complexation in microfluidic conditions:

$$\begin{cases} U\frac{\partial C_A}{\partial l} = D_A\frac{\partial^2 C_A}{\partial w^2} + D_A\frac{\partial^2 C_A}{\partial h^2} - k_1 C_A C_B + k_2 C_C \\ U\frac{\partial C_B}{\partial l} = D_B\frac{\partial^2 C_B}{\partial w^2} + D_A\frac{\partial^2 C_A}{\partial h^2} - k_1 C_A C_B + k_2 C_C \\ U\frac{\partial C_C}{\partial l} = D_C\frac{\partial^2 C_C}{\partial w^2} + D_A\frac{\partial^2 C_A}{\partial h^2} + k_1 C_A C_B - k_2 C_C \end{cases} \quad (5)$$

Where $U = f(w,h)U_{max}$ is the pressure-driven flow with parabolic profile.

Figure 2 demonstrates the results of 3D numerical simulations performed in Matlab software package for a polyelectrolyte-surfactant system. The association reaction occurs in the microchannel in the points where associating molecules meet by diffusion. The sandglass shape of the reaction front is the effect of the Hagen-Poiseuille flow profile in a rectangular microchannel: the flow velocity tends to zero near the walls.

4. Results and discussion

Figure 2 shows the results of 3D numerical simulations performed in Matlab PDE toolbox for a polyelectrolyte-surfactant system. The association reaction occurs in the microchannel in the points where associating molecules meet by diffusion. The sandglass shape of the reaction front is the effect of the Hagen-Poiseuille flow profile in a rectangular microchannel: the flow velocity tends to zero near the walls.

Polymer-surfactant systems demonstrate complex phase behavior depending on the ratio of components (the Z value). These ratios of surfactant and polymer concentrations Z were calculated by the ratio of the concentration of surfactant molecules in solution to the concentration of polyelectrolyte monomer units:

$$Z = \frac{C^{surfactant}}{C^{Polymer}_{mon}} \quad (6)$$

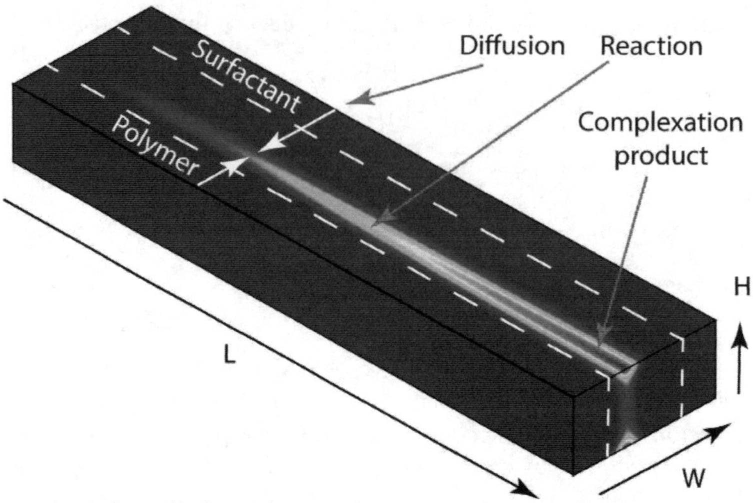

Figure 2. Formation of a polyelectrolyte-surfactant supramolecular system in a microchannel according to numerical 3D-simulation in Matlab PDE toolbox.

When the Z parameter reaches a certain value, the system undergoes phase separation and a line of precipitates emerge in a microchannel.

In bulk PAA-CTAB solutions with various ratios of components, formation of insoluble polymer-surfactant complexes and their precipitation occur at $Z<<0.1$. Further increase of surfactant-polyelectrolyte concentration ratio produces more turbid systems due to continuing phase separation. At $Z>0.7$-0.8, however, the turbidity of PAA-CTAB solutions decreases and precipitate dissolves at $Z\approx1$ due to their solubilization be excessive CTAB additives.

In microfluidic devices supramolecular systems can precipitate even if no such precipitation occurs in a macroscopic solution. Figure 3 demonstrates precipitates of PAA-CTAB complexes synthesized in a microchannel at $Z = 1$.

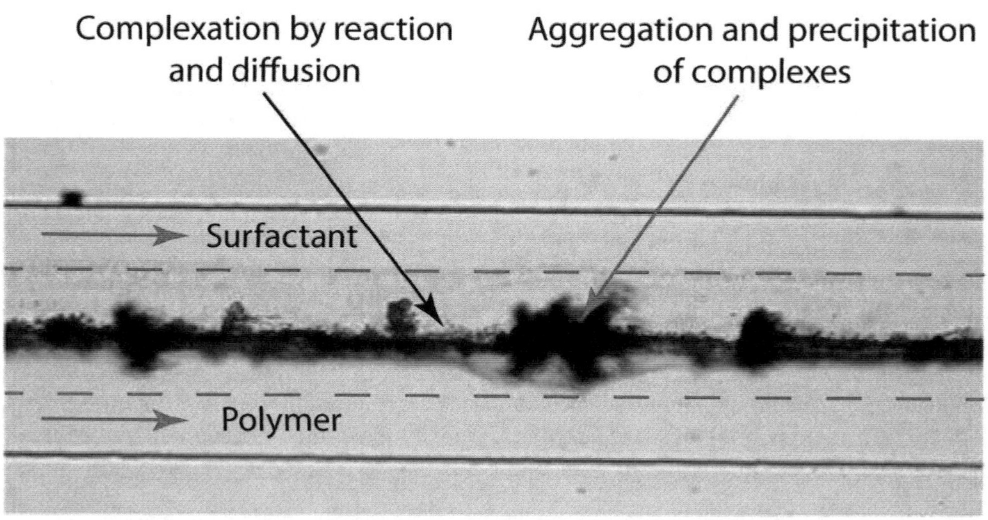

Figure 3. Optical microscopy photo of insoluble polymer-surfactant particles precipitated in a microchannel.

Location of precipitated particles correspond to the maximum of the product concentration calculated by numerical simulations and shown in Fig. 2.

IOP Conf. Series: Materials Science and Engineering **904** (2020) 012005 doi:10.1088/1757-899X/904/1/012005

Among various advantages of microfluidic methods, an ability to control properties of supramolecular systems by varying microchannel geometry and flowrates is important for fabrication of smart soft matter materials. In this work, we managed to control the size of PAA-CTAB aggregates by varying flowrates of reagents and the solvent (Figure 4).

The solutions of PAA-CTAB complexes were taken at the output of the 3-input microfluidic chip. Solutions contain particles in the range of ~170-310 nm depending on the synthesis conditions (flowrates). The smallest aggregates form and the largest aggregates form if the solvent flowrate is reduced to zero. Such a behavior corresponds to the properties of macroscopic polymer-surfactant solutions characterized in our previous works [19,20] as well as for the behavior of PAA-CTAB solutions we synthesized in bulk conditions.

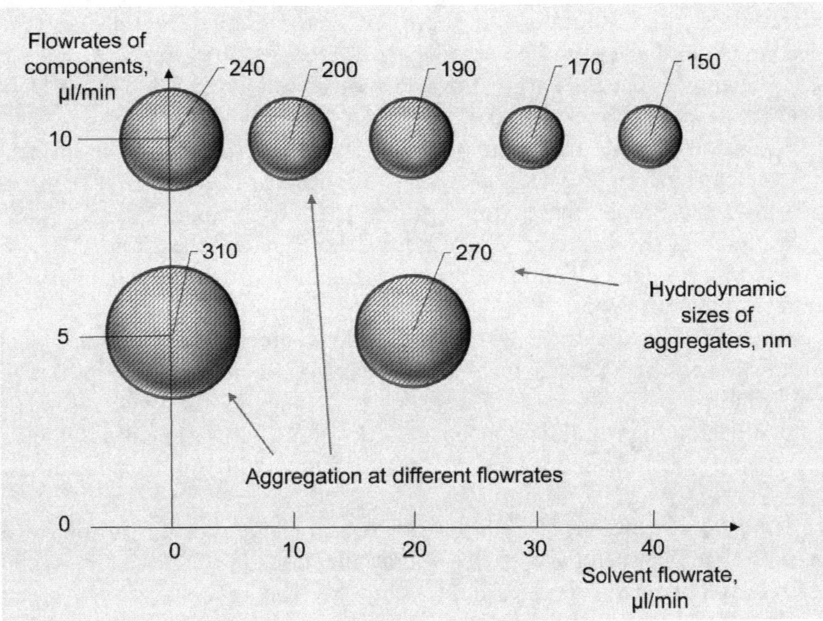

Figure 4. Hydrodynamic diameters of PAA-CTAB complexes synthesized in microfluidic experiments.

5. Conclusions

Microfluidics offers additional factors to control complexation of polyelectrolytes and surfactants. In addition to the ratio of components in bulk solutions, microfluidics offers such factors as flowrates of surfactant and polymer solutions and the solvent. The ratio of these flowrates contributes to association of polymer and surfactant molecules by diffusion and their subsequent aggregation.

Interaction of polymers and surfactants in microfluidic systems is governed by simultaneous convection, diffusion and kinetic processes. Numerical solutions of convection-diffusion-reaction equations provide a predictive model for association of polymers and surfactants in a microchannel. Parabolic profiles of pressure-driven microchannel flows are responsible for non-uniform distribution of the reaction product across the microchannel. Complexation reactions start at microchannel top and bottom sections and then proceed to other sections with the increase of microchannel length.

Complexation, aggregation and precipitation reactions occurring at microchannels with parallel flows of reacting supramolecular components such as polymers and surfactants contribute to fabrication of smart supramolecular structures that are sensitive to several factors. Their structural properties such as hydrodynamic size can be controlled by variation of flowrates od microchannel parameters in accordance with solution of the respective convection-diffusion-reaction equations.

6. Acknowledgments
This research was supported by the grant awarded to Kazan National Research Technological University by the Ministry of Higher Education and Science of the Russian Federation and dedicated to development of top academic products at regional higher education institutions in cooperation with the Russian universities at Top-200 of the global university rankings.

References
[1] Du A.W., Stenzel M.H. Drug Carriers for the Delivery of Therapeutic Peptides. Biomacromolecules. Volume 15(4), 2014, pp. 1097-1114.
[2] Stupp S.I., Palmer L.C. Supramolecular Chemistry and Self-Assembly in Organic Materials Design. Chemistry of Materials. Volume 26(1),2014, pp. 507-518.
[3] Borro B.C., Malmsten M. Complexation between Antimicrobial Peptides and Polyelectrolytes. Advances in Colloid and Interface Science. Volume 270, 2019, pp. 251-260.
[4] Hu Q., Li H., Wang L., Gu H., Fan C. DNA Nanotechnology-Enabled Drug Delivery Systems. Chemical Reviews. Volume 119(10), 2019, pp. 6459-6506.
[5] Bourganis V., Karamanidou T., Kammona O., Kiparissides C. Polyelectrolyte Complexes as Perspective Carriers for the Oral Delivery of Protein Therapeutics. European Journal of Pharmaceutics and Biopharmaceutics. Volume 111, 2017, pp. 44-60.
[6] Liu J.Y., Wang J.G., Li N., Hui Zh., Zhou H.J., Sun P.Ch., Chen T.H. Polyelectrolyte–Surfactant Complex as a Template for the Synthesis of Zeolites with Intracrystalline Mesopores. Langmuir. Volume 28, 2012, 28, pp. 8600-8607.
[7] Ma J., Lee S.M.Y., Yi C., Li C.W. Controllable synthesis of functional nanoparticles by microfluidic platforms for biomedical applications – a review. Lab Chip. 2017, 17, 209-226.
[8] Martins C., Araújo F., Gomes M. J., Fernandes C., Nunes R., Li W., Santos H.A., Borges F., Sarmento B. Using microfluidic platforms to develop CNS-targeted polymeric nanoparticles for HIV therapy. European Journal of Pharmaceutics and Biopharmaceutics. Volume 139, 2019, pp. 111-124.
[9] Maeki M., Kimura N., Sato Y., Harashima H., Tokeshi M. Advances in microfluidics for lipid nanoparticles and extracellular vesicles and applications in drug delivery systems. Advanced Drug Delivery Reviews. Volume 128, 2018, pp. 84–100.
[10] Zhang L., Chen Q., Ma Y., Sun J. Microfluidic Methods for Fabrication and Engineering of Nanoparticle Drug Delivery Systems. ACS Applied Bio Materials. Volume 3(1), 2020, pp. 107-120.
[11] Iliescu C., Marculescu C., Venkataraman S., Languille B., Yu H., Tresset. G. On-Chip Controlled Surfactant–DNA Coil–Globule Transition by Rapid Solvent Exchange Using Hydrodynamic Flow Focusing. Langmuir. Volume 30(44), 2014, pp. 13125–13136.
[12] Tresset G., Marculescu C., Salonen A., Ni M., Iliescu C. Fine Control Over the Size of Surfactant–Polyelectrolyte Nanoparticles by Hydrodynamic Flow Focusing. Analytical Chemistry. Volume 85(12), 2013, pp. 5850–5856.
[13] Ho Y.-P., Grigsby C. L., Zhao F., Leong. K. W. Tuning Physical Properties of Nanocomplexes through Microfluidics-Assisted Confinement. Nano Letters. Volume 11(5), 2011, pp. 2178–2182.
[14] Chiesa. E., Dorati. R., Pisani. S., Conti B., Bergamini G., Modena T., Genta I. The Microfluidic Technique and the Manufacturing of Polysaccharide Nanoparticles. Pharmaceutics. Volume 10(4), 2018, p. 267.
[15] Iliescu C., Tresset G. Microfluidics-Driven Strategy for Size-Controlled DNA Compaction by Slow Diffusion through Water Stream. Chemistry of Materials. Volume 27(24), 2015, pp. 8193–8197.
[16] Berthier J., Silberzan P. Microfluidics for Biotechnology. Second Edition. Artech House. 2009, 483 p.
[17] Tabeling P. Introduction to Microfluidics. Oxford University Press. 2010, 310 p.

[18] Bruus H. Theoretical Microfluidics. Oxford Master Series in Physics (Book 18). Oxford University Press. 2007, 288 p.

[19] Shilova S. V., Tret'yakova A. Ya., Bezrukov A. N., Myagchenkov V.A. Barabanov V. P. Effect of molecular weight of poly-N-benzyl-N,N-dimethyl-N- methacryloyloxyethylammonium chloride on its complexation with sodium dodecyl sulfate. Russian Journal of Applied Chemistry. Volume 80(9), 2007, pp. 1579-1584. doi: 10.1134/S1070427207090261.

[20] Shilova S. V., Bezrukov A. N., Tret'yakova A. Y., Voronin M. A., Zakharova L. Y., Barabanov V. P. Effect of the length of surfactant hydrocarbon radicals on the association of cationic polyelectrolytes with alkyl sulfates in water-alcohol solutions. Polymer Science Series A. Volume 54(1), 2012, pp. 19–25. doi: 10.1134/S0965545X12010075.

IOP Conf. Series: Materials Science and Engineering **904** (2020) 012006 doi:10.1088/1757-899X/904/1/012006

Improving the start-up parameters of a gas turbine engine by adaptation the air turbine starter

Grigorii Popov, Valery Matveev, Vasilii Zubanov, Yulia Novikova, Daria Kolmakova and Anastasia Korneeva

Samara National Research University, Samara, 443086, Russia

grishatty@gmail.com

Abstract. The paper describes the process of optimizing the blades of a starter air turbine for a gas turbine engine launch system. This task was initiated by the necessity to use an existing turbine when starting a new engine. During the study, it was found that the turbine, although it ensures the conditions for the joint operation with the auxiliary power unit turbine and meet the strength constraints, it does not allow to start the engine within the required time. As a result of studies using mathematical optimization methods involving commercial programs, a variant was found to modernize the baseline turbine, which provides an acceptable value of torque on the shaft with an adequate start-up time at all operating modes with minimal design changes.

1. Nomenclature

GTE = gas turbine engine
APU = auxiliary power unit
NGV = nozzle guide vane
ATS = air turbo starter
IGV = inlet guide vane
K_G = mass flow parameter
K_N = power parameter
P = pressure
T = temperature
G = mass flow rate
α_{IGV} = IGV stagger angle
π = pressure ratio
ΔG = losses (of air flow in the aircraft
Δp^* = total pressure losses in the aircraft ducts
ΔT^* = total temperature losses in the aircraft ducts
N = power
p_h = ambient pressure
t_h = ambient temperature
n = rotational speed
M = torque
τ = time

2. Introduction

The start-up of an aircraft gas turbine engine is an important mode that largely determines the safety, operational efficiency and reliability of the engine and the entire aircraft. The gas turbine engine start-up system includes a set of various devices and units: a starter, auxiliary power unit, air and fuel communications, automatic control system, transmission, power supply system, ignition system, etc. For reliable engine starting, the operation of all these systems must be consistent with each other [1].

Currently, the civil aircraft engine starting system is often built based on an air turbine mechanically connected to the rotor of the gas turbine engine, receiving compressed air from the APU compressor (Figure 1).

The group of the paper's authors is employed by the Department of Aircraft Engine Theory of Samara National Research University [2] and has extensive experience in studying and improving the working process of various components of a gas turbine engine [3, 4]. Currently, the scientific group takes part in joint work in the interests of various enterprises that are the part of the United Engine Corporation (Russia) [5]. In particular, one of the Customers set the task of assessing the feasibility of using an air turbostarter manufactured at the enterprise to launch a new turbofan engine of the same class. The design of the considered ATS is shown in Figure 2.

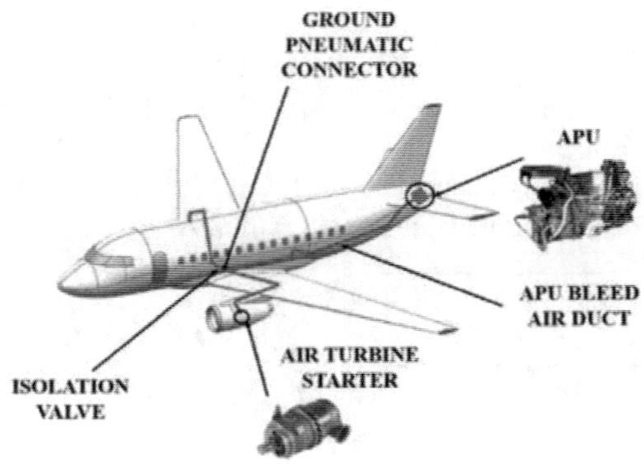

Figure 1. Concept scheme of the starting system with an air turbine.

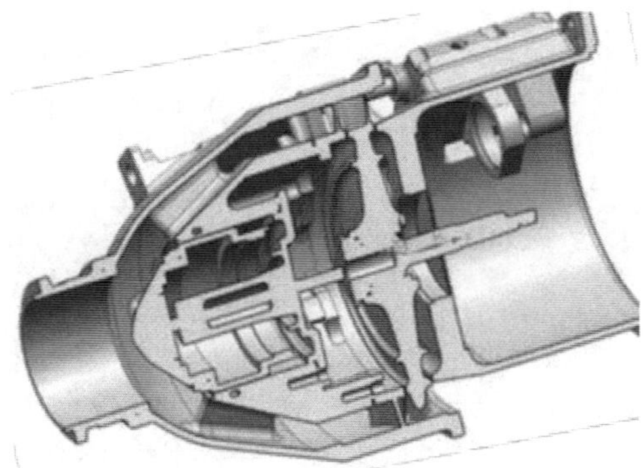

Figure 2. Design of the investigated starter air turbine.

The team was assigned the following tasks:

1) the ATS operation must be coordinated with the operation of the APU at all operating modes (under various atmospheric conditions, flight speeds and altitudes);
2) startup time must be minimized (the minimum allowed time was set);
3) the torque on the output shaft must not exceed the maximum value according to the strength conditions of the reduction gearbox and gear box of the engine drives.

In other words, the main task was to modify the flow part of the existing ATS so that it would satisfy the above conditions when working in a modified launch system (another gas turbine engine and APU). Moreover, the design of the baseline turbine must be kept as much as possible to reduce costs. Thus, the task is a typical optimization problem in which the turbine geometry must be changed (considering various constraints: structural and strength) to improve the required criterion (start-up time).

In open scientific and technical sources, there are many publications on the modernization and optimization of the turbine workflow [6-8]. However, all of them are aimed at improving the efficiency of turbines or their reliability. In current case, the unusual optimization criteria take place and it is necessary to check the condition for the joint operation of the ATS and APU and to calculate the start-up time of the gas turbine engine during optimization.

To solve the problem, an algorithm was developed for matching the working process of the APU and the air turbine used to start the engine. This method was described in detail in [9].

The essence of the method is that the characteristics of the APU and ATS are converted to a general form of the dependence of the reduced flow parameter K_G on the expansion ratio of air in the turbine $K_{G\ ATS} = f(\pi_{ATS})$. The intersection of the lines of the APU and turbine operating modes suggests that the conditions for joint work are fulfilled at these modes. At the intersection points of the characteristics $K_{G\ ATS} = f(\pi_{ATS})$ and $K_{G\ APU} = f(\pi_{APU})$, the parameters at the APU outlet $p^*_{APUoutlet}$, $T^*_{APUoutlet}$, $G_{APUoutlet}$ are calculated and with their help the characteristics of the turbine (in particular, the dependence of power on rotational speed) are determined under the found conditions. Then, using the power characteristics of the ATS, the start time of the gas turbine engine is determined. The algorithm for its calculation is described in [9]. The conformity of the maximum torque to the strength constraints is also checked.

Thus, based on the intersection points of the above characteristics, the physical characteristics of the ATS are found when working together with the APU at all its modes.

Under the flow parameter K_G is meant a physical quantity that is calculated using the following expression:

$$K_G = \frac{G_{ATSinput} \cdot \sqrt{T^*_{ATSinput}}}{p^*_{ATSinput}}, \tag{1}$$

where $G_{ATSinput}$ - the value of air mass flow rate through the ATS;

$T^*_{ATSinput}$ - the value of the total temperature at the ATS inlet;

$p^*_{ATSinput}$ - the value of the total pressure at the ATS inlet.

When calculating the flow parameter K_G for the APU, leakages and pressure and temperature losses in the pipelines of the launch system on the way to the turbine are considered.

$$K_{G\,\mathrm{APU}} = \frac{(G_{\mathrm{APUoutlet}} - \Delta\mathrm{G}) \cdot \sqrt{(T^*_{\mathrm{APUoutlet}} - \Delta T^*)}}{p^*_{\mathrm{APUoutlet}} - \Delta p^*_{hydr.}}, \tag{2}$$

$$\pi_{\mathrm{APU}} = \frac{p^*_{\mathrm{APUoutlet}} - \Delta p^*_{hydr.}}{p_h}, \tag{3}$$

where $K_{G\,\mathrm{APU}}$ – APU air mass flow parameter determined at the inlet of the ATS considering losses in the pipelines;

π_{APU} – pressure ratio in the ATS determined by the parameters of the air at the inlet to the ATS considering losses in the pipelines;

G_{ATSinput} – the value of the mass flow rate of air taken from the APU considering losses in the pipelines, $\left[\frac{kg}{s}\right]$:

$$G_{\mathrm{ATSinput}} = G_{\mathrm{APUoutlet}} - \Delta\mathrm{G} \tag{4}$$

T^*_{ATSinput} – the value of the temperature of air taken from the APU considering losses in the pipelines, [K]:

$$T^*_{\mathrm{ATSinput}} = T^*_{\mathrm{APUoutlet}} - \Delta T^*, \tag{5}$$

p^*_{ATSinput} – the value of the total pressure of air taken from the APU considering losses in the pipelines, $[\frac{kgf}{cm^2}]$:

$$p^*_{\mathrm{ATSinput}} = p^*_{\mathrm{APUoutlet}} - \Delta p^*_{\mathrm{hydr.}}, \tag{6}$$

p_h – atmospheric pressure for altitude in standard atmospheric conditions, $\left[\frac{kgf}{cm^2}\right]$.

Values of losses of mass flow rate, temperature and pressure in the system are set by the Customer according to the experience of operating an engine of a similar class. The dependence $K_{G\,\mathrm{ATS}} = f(\pi_{\mathrm{ATS}})$ obtained for the used APU is shown in Figure 3.

Characteristics of the turbine of the form $K_{G\,\mathrm{ATS}} = f(\pi_{\mathrm{ATS}})$ (Figure 4) were obtained by CFD modeling of the working process in the investigated turbine (Figure 2). Its mathematical model was created using NUMECA software system. The model included a structural grid of finite volumes of 92 thousand elements. The value of the parameter y+ did not exceed 1. The Spalart-Allmaras turbulence model was used. In general, the settings of the numerical model of an air turbine corresponded to the typical settings used in the calculation of the working process in turbines [10].

The calculation results obtained using the created calculation model were compared with the experimental data provided by the customer (Figure 5). The relative error in determining the ATS power was less than 2.5%, and the air mass flow rate was less than 2%.

The obtained characteristics $K_G = f(\pi_{ATS})$ for the ATS and APU were combined at one diagram (Figure 6).

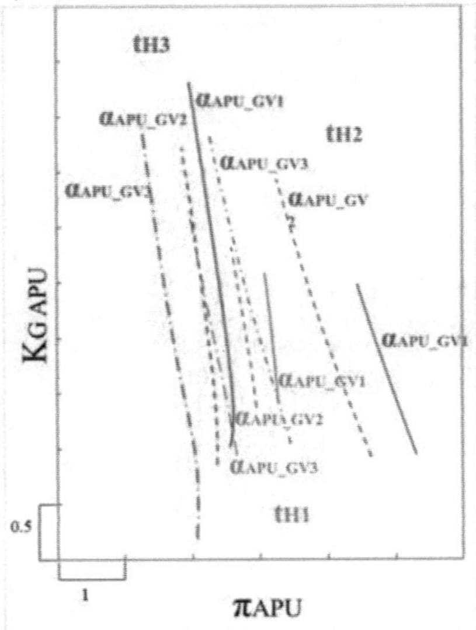

Figure 3. Characteristics $K_{G\ APU} = f(\pi_{APU})$ of used APU.

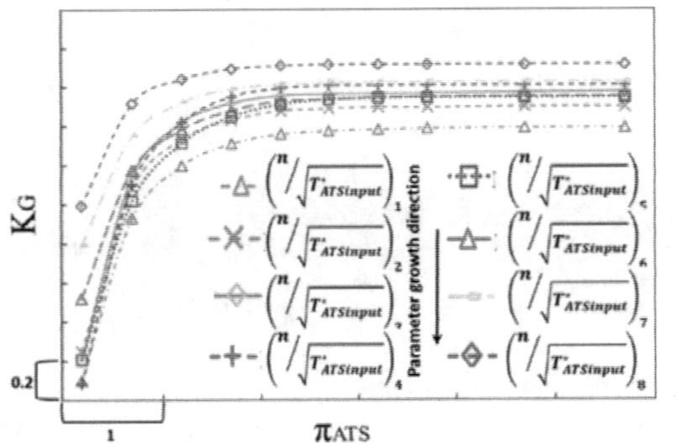

Figure 4. Mass flow characteristics $K_{G\ ATS} = f(\pi_{ATS})$ of investigated air turbine.

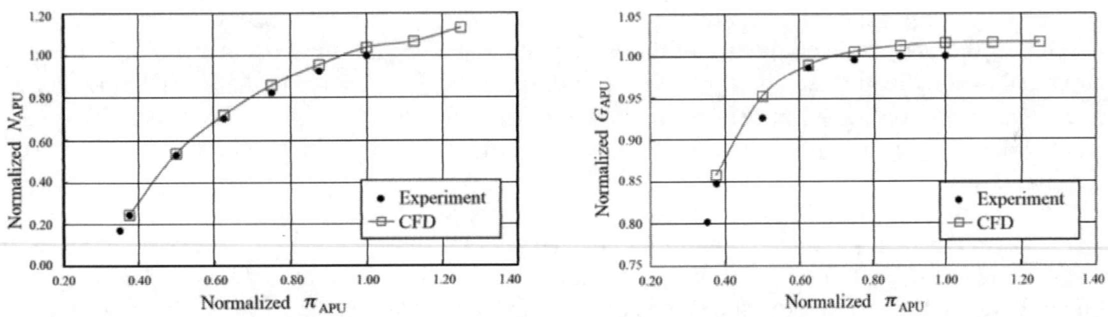

Power characteristics Mass flow characteristics

Figure 5. Comparison of calculation results obtained with the created computational model with experimental data.

As can be seen from Figure 6, all the lines of APU and ATS operating modes intersect, which indicates that the condition for their joint operation is fulfilled at all operating modes. However, it was found during the calculation, that the start-up time of the gas turbine when using the ATS exceeds the value specified in the technical requirements by 12.6%. At the same time, the maximum torque on the turbine shaft did not exceed the permissible value. Thus, it is necessary to increase the turbine power (without exceeding its limit value) and the air flow through it in order to reduce the start-up time to use the turbine for a new GTE start system.

Analysis of the turbine working process showed that the required increase in the turbine power and mass flow rate can be achieved with minimal changes in the initial design only by modernizing the shape of the nozzle guide vanes. In the other words, the modernization of the existing ATS to the new requirements can be performed while substantially preserving the design. In fact, only one part will be changed that is the nozzle block.

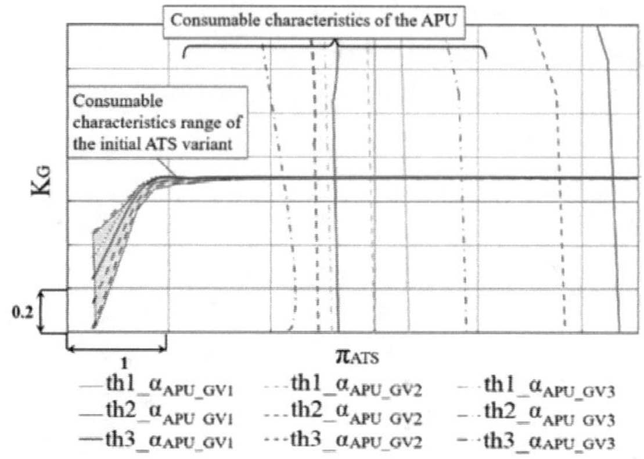

Figure 6. Combined flow characteristic of the APU and ATS.

3. Parametric studies

As noted above, it is necessary to increase the turbine power in order to meet the requirements for starting time. This can be achieved by increasing the flow rate through the turbine. The element limiting the flow in the turbine is NGV. Therefore, the flow rate can be increased in two ways: by changing the stagger angle and by trimming the trailing edge.

To search for a possible solution, parametric studies were conducted on the influence of these parameters on the startup time.

3.1. Changing the NGV stagger angle

During the study of the influence of the stagger angle of NGV on the flow rate of the working fluid and the power of the turbine, the angle varied in the range of +2° relative to the initial value. In total, five points were considered (including the initial geometry) with a step of 0.5°. A larger change in the angle will lead to a change in the design of adjacent parts and a significant amount of alteration of the original design.

The obtained dependences of the mass flow parameter K_G, start-up time and torque on changes in NGV stagger angle are shown in Figure 7. It can be seen that with a maximum increase in the NGV stagger angle, the start-up time of the gas turbine engine was reduced by only 7.6%. This value is 5.7% higher than the limit value (i.e., such a turbine does not satisfy the requirements). The maximum value of torque was 27.5% less than the maximum allowable, which meets the requirements of the technical specifications. Thus, changing only the stagger angle within the existing constraints on the axial length of the NGV part is not enough to increase the turbine power and to reduce the start-up time of the GTE.

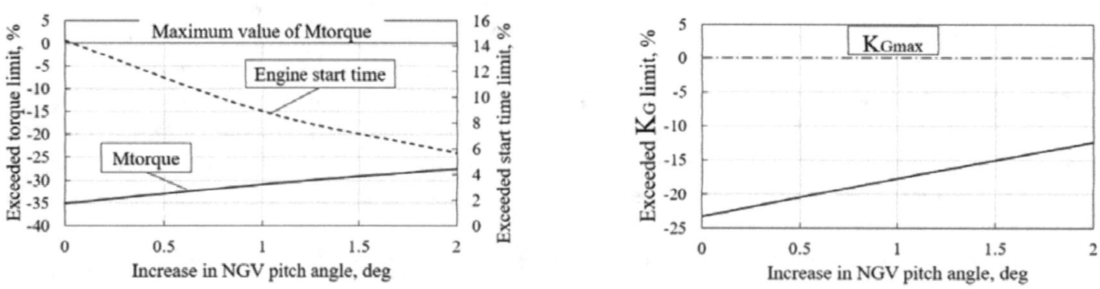

Figure 7. Dependences of the parameters of the starter turbine and the start-up time of the gas turbine engine on the stagger angle of the ATS NGV.

3.2. *Trimming the trailing edge of NGV*

Five variants for trimming the trailing edge of the NGV were considered (Figure 8). The dependences of the flow rate parameter K_G, start-up time and torque on the value of trimming the NGV trailing edge are obtained. They are shown in Figure 8. It can be seen from the presented results that, the start-up time was reduced by 10.7% by trimming the NGV trailing edge, which is 2.1% more than the maximum allowable τ_max. The maximum torque value was 21.8% less than the maximum allowable M_{torque_max}.

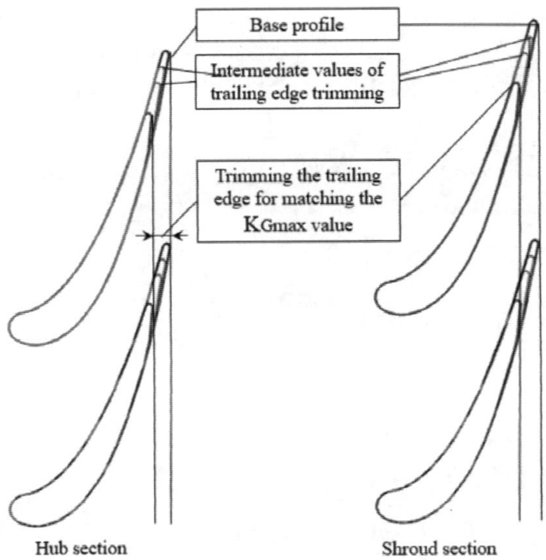

Figure 8. The hub and shroud sections of the NGV variants with trimming trailing edge.

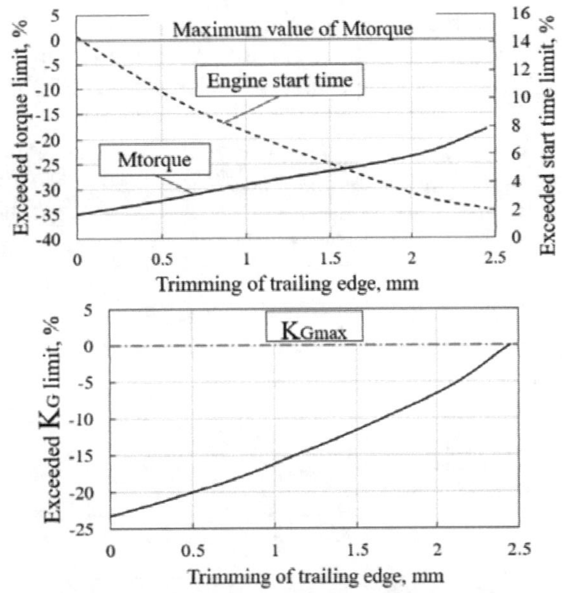

Figure 9. Dependences of the mass flow parameter K_G, start-up time and torque on the value of trimming the NGV trailing edge.

Thus, trimming the NGV trailing edge also does not provide the GTE start-up time specified in the technical task. During the study, it was found that the proposed modernization options cause significant deviations of the inlet flow angles for the rotor blade, which are the reason of significant dynamic stresses in it.

Thus, none of the considered variants of ATS modernization satisfies the requirements of the technical specifications with its design is preserved to the maximum. For this reason, it was decided to use mathematical optimization methods to find an acceptable variant.

4. Optimization Algorithm for Starter Air Turbine

A further search for a new shape of ATS NGV was carried out using multicriteria mathematical optimization methods. For this, an algorithm was developed that is built around the IOSO optimizer program [11]. The choice of the IOSO program is due to the large number of its successful application in the tasks of aircraft engine field, including by the authors of the paper [12-17]. The specified program was used as a finished commercial product. No upgrades were made to optimization algorithms. A description of the algorithms used in the program can be found on the website and publications of the program developer.

The schematic diagram of the developed ATS optimization algorithm is shown in Figure 10. At each optimization step, the *IOSO PM* optimizer generates a vector of input parameters $x_1, x_2, x_3, \ldots, x_n$. The values of the input parameters describe the geometry of the nozzle guide vanes in a parametric form. The vector of variable parameters is transferred to the reprofiling unit, in which the *Numeca AutoBlade* program [18] converts the blades and saves them as geometry files in the *.*GeomTurbo*

IOP Conf. Series: Materials Science and Engineering **904** (2020) 012006 doi:10.1088/1757-899X/904/1/012006

format. Then, *Numeca AutoBlade 5* creates a mesh model using new blades. At the next step, CFD calculation is performed with the new mesh model. Processing the results of CFD modeling is performed by a special script using *Numeca FineTurbo* [19] and small applications from the NET Framework library. As a result, several output files are created containing the ATS operation parameters of interest in the text format. Then these parameters are transferred to the IOSO optimizer, where the results are processed and the current optimal variant of the ATS NGV is selected, and a new set of input parameters $x_1, x_2, x_3, \ldots, x_n$. is created. This process is iterative and runs until the desired extremum is found, taking into account the given constraints.

The goal of optimizing the flow part of the ATS is to reduce the start-up time of a gas turbine engine while reducing the torque on the turbine output shaft (in order to increase the reliability). The presence of two optimization criteria ensures more stable operation of the IOSO optimizer [20].

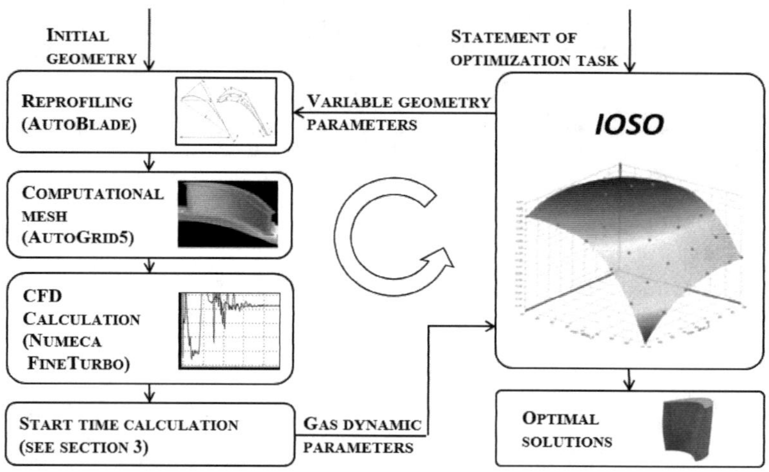

Figure 10. Schematic diagram of the applied methodology for multicriteria optimization of ATS.

Based on the foregoing, the following criteria were selected for optimization:

1) GTE start time (it must be reduced);
2) the torque on the turbine shaft (it must be reduced).

The following parameters were set as constraints:

1) the torque at all modes must be less than the maximum allowable value;
2) the flow parameter K_G must be within the range $K_{G\_min} \ldots K_{G\_max}$, where the value K_{G_min} corresponds to the minimum value of K_G from all operating modes of the APU, and K_{G_max} - to the maximum value of Kg from all operating modes of the APU. This requirement is necessary to fulfill the conditions for the joint work of the APU and the ATS at all operating modes.

The parametrization scheme for the sections of the ATS nozzle guide vanes is shown in Figure 11. Changing the geometry of the first nozzle guide vanes was carried out in two control sections (hub and shroud). Each section was described by 14 independent variables (chord, inlet and outlet design angles, profile stagger angle, edge radii, position of control points of the spline of the pressure side and suction side, etc.). In total, 28 variables were used to describe the geometry of the ATS nozzle guide vanes.

Such a statement of the optimization problem makes it possible to find the Pareto front of optimal solutions in which each value of the gas turbine engine start-up time corresponds to the minimum torque at which it can be achieved (Figure 12). Then, in accordance with the task requirements, a variant is selected that provides the minimum start-up time, that can be achieved with a torque value acceptable under the conditions of transmission strength. Obviously, the requirement to reduce the start-up time will shift the selected point on the Pareto front to the left as far as possible, as much as the torque limitations allow.

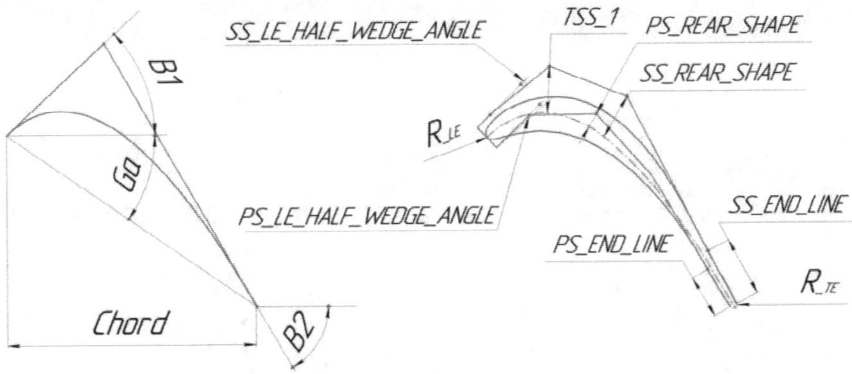

Figure 11. Parametrization scheme for the sections of the ATS nozzle guide vanes.

5. Optimization Results

The optimization process was stopped after 200 calculation cycles. As a result, the Pareto front was obtained between the values of the start-up time of the gas turbine engine and the torque on the output shaft of the ATS (Figure 12).

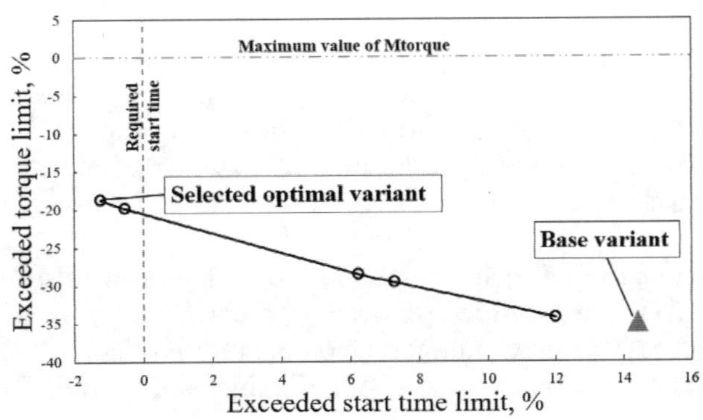

Figure 12. Pareto Front "Torque - Engine Start-Up Time" of the ATS variant at the most loaded operating mode.

For the current problem, the case with maximum torque and minimum start-up time is of the greatest practical interest. The start-up time of the gas turbine engine for the selected variant decreased by 13.7% relative to the variant with the baseline ATS geometry and was 1.3% less than the maximum permissible according to the technical requirements. The maximum torque among all the matched operating modes of the APU did not exceed the maximum allowable M_{torque_max} by 18.5%. The value of the mass flow parameter K_G remained within the specified limits $K_{G\_min}...K_{G\_max}$ and was less than the value K_{G_max} by 3.8%. The latter circumstance ensured the preservation of the coordinated work of the APU and ATS at all operating modes. Thus, the main practical goal of the study was achieved.

Figure 13 shows a comparison of the NGV cross-sections of the optimized and baseline ATS variants. The increase in the cross-sectional area and, accordingly, the flow parameter K_G, occurred due to a decrease in the blade thickness, and due to a decrease in the chord of the blade, especially in the shroud section. Since the shape of only the stator blades was adjusted, the strength of the most loaded turbine elements did not change significantly.

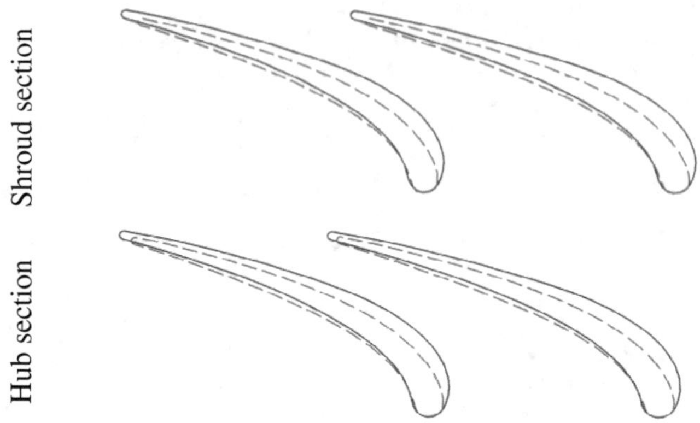

Figure 13. Blade-to-blade sections of ATS NGV: solid line - the baseline variant; dashed line - optimized variant.

6. Conclusion

This paper presents the results of the modernization of the flow part of an existing air turbine for starting system for application at another engine of the same class. It was shown that the initial turbine satisfies the requirements of working together with the APU at all operating conditions, provides satisfaction of strength requirements, but the start-up time when using it exceeds by 12% the maximum time required in the technical specifications. During the analysis, it was found that it is necessary to increase the power of the turbine by increasing air flow rate in order to fulfill the requirements. Changing the parameters of the turbine was carried out by changing the shape of the NGV. Conducted parametric research did not allow to find a solution. Therefore, the problem was solved using the original algorithm using a commercial optimizer program. As a result, a solution was found that made it possible to find such a configuration of the turbine flow path only by changing one element (NGV unit) that ensures that the technical requirements in terms of starting time are met with a margin of 2%. In this case, the maximum torque on the turbine shaft is 18% less than the permissible value and the conditions for the joint operation of the turbine and the APU are fulfilled at all operating modes.

The authors hope that the selected air turbine will be tested with a new engine in the near future in order to make sure the conclusions are correct.

References

[1] Inozemcev, A. A., Nihamkin, M. A., Sandrackii, V. L. *Fundamentals of designing aircraft engines and power plants: textbook*. Moscow: Mashinostroenie, 2008. V.2. 368 p.

[2] Samara University, Access mode: https://ssau.ru/english/

[3] Marchukov, E., Egorov, I., Popov, G., Baturin, O., Goriachkin, E., Novikova, Y. and Kolmakova, D. "Improving of the working process of axial compressors of gas turbine engines by using an optimization method", *IOP Conference Series: Materials Science and Engineering*. Vol. 232. No.1 (2017): Article number 012041. DOI 10.1088/1757-899X/232/1/012041

[4] Matveev, V., Baturin, O. and Popov, G. "The Optimization of Four-Stage Low Pressure Turbine with Outlet Guide Vane", *IOP Conference Series: Materials Science and Engineering*. Vol. 302. No.1 (2018): Article number 012037. DOI 10.1088/1757-899X/302/1/012027

[5] United engine corporation, Access mode: https://www.uecrus.com/eng/

[6] Marchukov, E.Yu., Egorov, I.N. "Gas Dynamic Modernization of Axial Uncooled Turbine by Means of CFD and Optimization Software", *IOP Conference Series: Materials Science and Engineering*. Vol. 302, No. 1 (2018): Article number 012027. DOI 10.1088/1757-899X/232/1/012041

[7] Châtel, A., Verstraete, T., Coussement, G. "Multipoint optimization of an axial turbine cascade using a hybrid algorithm", *Proceedings of the ASME Turbo Expo 2019: Turbomachinery Technical Conference and Exposition*, Phoenix, United States. Article number GT2019-91471. DOI 10.1115/GT2019-91471.

[8] Asgarshamsi, A., Hajilouy-Benisi, A., Assempour, A., Pourfarzaneh, H. "Multi-point optimization of lean and sweep angles for stator and rotor blades of an axial turbine", *Proceedings of the ASME Turbo Expo 2014: Turbine Technical Conference and Exposition*, Dusseldorf, Germany. Article number GT2014-27016. DOI 10.1115/GT2014-27016.

[9] Zubanov, V.M., Matveev, V.N., Popov, G.M., Novikova, Y.D. "Maturation of an aircraft gas turbine engine air turbo start unit on the basis of its joint action with auxiliary power plant". *VESTNIK RGATU imeni P.A. Solov'eva*, No. 4 (51) 2019, pp. 27-32.

[10] Popov, G., Matveev, V., Baturin, O., Novikova, J., Kolmakova, D. and Volkov, A. "Selection of Parameters for 3D Finite-volume Mesh for CFD Simulation of Axial Turbines", *MATEC Web of Conferences*. Vol. 220 (2018): Article number 07001. DOI 10.1051/matecconf/201822007001.

[11] IOSO Optimization Technology. Access mode: http://www.iosotech.com

[12] Marchukov, E., Egorov, I., Popov, G., Baturin, O., Goriachkin, E. and Novikova, Y. "Optimization of a three spool axial compressor to increase the efficiency of a gas turbine engine", *IOP Conference Series: Materials Science and Engineering*. Vol. 604. No. 1 (2019): Article number 012048. DOI 10.1088/1757-899X/604/1/012048.

[13] Marchukov, E., Egorov, I., Kretinin, G., Karonic, B. & Fedechkin, K. 2019, "Optimization of geometry blade for modern high pressure compressor", *IOP Conference Series: Materials Science and Engineering*. Vol. 604. No. 1 (2019): Article number 012049. DOI 0.1088/1757-899X/604/1/012049.

[14] Marchukov, E.Y. and Egorov, I.N. "Gas Dynamic Modernization of Axial Uncooled Turbine by Means of CFD and Optimization Software", *IOP Conference Series: Materials Science and Engineering*. Vol. 302. No. 1 (2018): Article number 012027. DOI 10.1088/1757-899X/302/1/012027

[15] Marchukov, E.Y., Egorov, I., Popov, G., Salnikov, A., Goriachkin, E. and Kolmakova, D. "Multidisciplinary optimization of the working process of uncooled axial turbine according to efficiency and strength criteria". *Proceedings of the ASME Turbo Expo*. GT2017-64843. Charlotte, United States. June 26-30, 2017. DOI 10.1115/GT2017-64843.

[16] Salnikov, A. And Danilov, M. "Multidisciplinary design optimization of a bladed disc for small-size gas-turbine engines". *Proceedings of the ASME Turbo Expo 2019*. GT2019-90974. Phoenix; United States; June 17-21, 2019. DOI: 10.1115/GT2019-90974

[17] Buyukli, T., Salnikov, A. and Fedorchenko, Y. "High-loaded compressor blisk-type impeller multidisciplinary optimization". *Proceedings of the ASME Turbo Expo 2017*. GT2017-64116. Charlotte, United States. June 26-30, 2017.DOI:10.1115/GT2017-64116

[18] Numeca AutoBlade, Access mode: https://www.numeca.com/product/finedesign3d

[19] Numeca FineTurbo, Access mode: https://www.numeca.com/product/fineturbo

[20] Kuzmenko, M.L., Shmotin, Yu.N., Egorov, I.N. and Fedechkin, K.S. "Optimization of the gas turbine engine parts using methods of numerical simulation." *Proceedings of the ASME Turbo Expo 2007*. GT2007-28205: pp. 425-431. Montreal, Canada. May 14-17, 2007. DOI 10.1115/GT2007-28205

IOP Conf. Series: Materials Science and Engineering **904** (2020) 012007 doi:10.1088/1757-899X/904/1/012007

Study on connection between shape of wheel and resistance force from ground for small wheel typed rovers

Tomohiro WATANABE and Kojiro IIZUKA

Shibaura Institute of Technology, 307 Fukasaku, Minuma-ku, Saitama City, Saitama, Japan, 337-8570

nb19109@shibaura-it.ac.jp

Abstract. In recent years, the small wheel typed rovers are focused on as an exploring rover. However, running performance of the small wheel typed rovers is not good. It is difficult to compact the ground by wheels because weight of these rovers is small. Therefore, drawbar pull is low. In previous research, the movement method which improves the running performance of these rovers using effects which are given by vibration to the ground was proposed. Some experiments were carried out in order to confirm effective of proposed method. From theses experimental results, we could get knowledge that how to use vibration to improve running performance for the small wheel typed rovers. Moreover, it was confirmed that sinkage of the wheels to the ground is increased when vibration is given to the ground from the wheel. Sinkage of the wheels to the ground decrease running performance of rover. In this research, the relationship between the shape of wheel and resistance force from the ground was confirmed in order to propose the wheel which reduce subsidence amount to the ground when vibration is given to the ground. In experimental results, the wheel which has flange on side reduced subsidence amount to the ground.

1. Introduction

In recent years, the small wheel typed rovers are focused on as an exploring rover[1][2][3]. Volume and weight of these rovers are small. Cost of exploring the moon and Mars can be reduced by using these rovers. However, running performance of the small wheel typed rovers is not good. It is difficult to compact the ground by wheels because weight of these rovers is small. Therefore, drawbar pull of them is low. In previous research, the movement method which improves the running performance of these rovers using effects which are given by vibration to the ground was proposed. When the loose ground is given vibration, there are two effects. One is to increase density and shear strength by compaction of the loose ground. The other is to increase sinkage of wheels because the ground is softened by moving particle of the ground. As a result, the value of shearing strength and sinkage of wheel were increased after vibration. Furthermore, these effects given by vibration are confirmed using the small wheel typed testbed. From theses experimental results, we could get knowledge that how to use vibration to improve running performance for the small wheel typed rovers.

Moreover, it was confirmed that sinkage of the wheels to the ground is increased when vibration is given to the ground from the wheel. Sinkage of the wheels to the ground decrease running performance of the rover. In this research, the relationship between the shape of wheel and resistance force from the ground was confirmed in order to propose the wheel which reduce subsidence amount to the ground when vibration is given to the ground.

2. Running method used vibration

In this chapter, mechanism of running method used vibration which is proposed by our study group is explained. In addition, how to use changes of the ground when giving vibration in proposed running method is explained.

The proposed wheel which gives the ground to vibration is shown in Fig. 1. The vibration motor is attached to internal of wheel. In previous study, the testbed was constructed by four vibration wheels. Moreover, it was confirmed that running performance of the testbed is improved by vibrating the front wheels and running. The ground is compacted by vibrating the front wheels. Therefore, it is considered that the traction force is increased because the rear wheels run on the compacted ground. Actually, it is confirmed that shear strength of the ground is increased by giving the ground to vibration[4]. Particle of the ground is moved by giving the ground to vibration as shown in Fig. 2. Shear strength of the ground is increased by moving particle of the ground.

However, sinkage of the wheels to the ground is also increased by moving particle of the ground. Sinkage of the wheels to the ground decrease running performance of the rover. Therefore, it is needed to develop the solution which reduce subsidence amount to the ground when vibration is given to the ground.

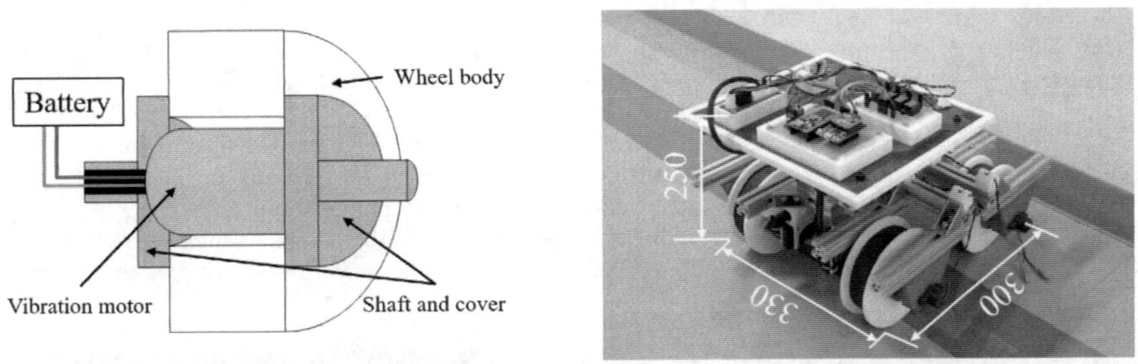

(a) Structure of vibration wheel (b) Overview of testbed

Figure 1. Overview of vibration wheel and testbed.

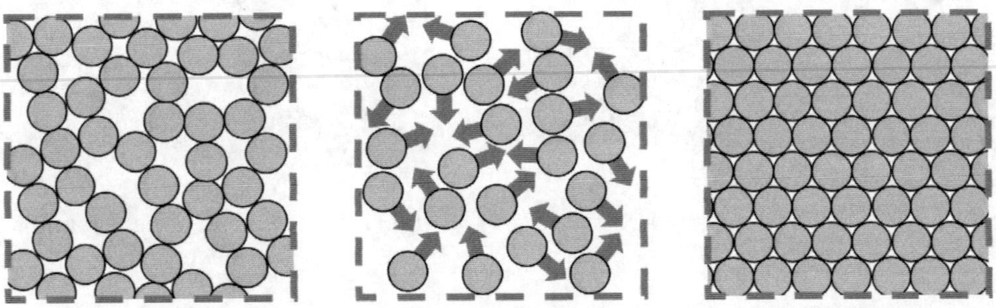

(a) Before vibration (b) Vibrating (c) After vibration

Figure 2. Movement particles which constitute ground when vibration is given to ground.

3. Penetration experiment for measuring resistance force from ground

Sinkage of the wheels to the ground is increased when resistance force from the ground is small. It is needed to confirm that relationship between the shape of wheel and resistance force from the ground. Therefore, the penetration experiment for measuring resistance force from the ground is conducted. In this chapter, the testing machine, the experimental method and the result of this experiment are reported.

3.1. Penetration testing machine

The penetration testing machine is explained. This machine is consisted of soil tank part, wheel part and actuator part as shown in Fig. 3. Fig. 4 shows movement of this machine. Wheel part is penetrated in soil tank part by actuator part. Wheel part is consisted of force sensor and vibration wheel. The resistance force from the ground when penetrating vibration wheel is measured by force sensor.

3.2. Content of experimental method

Flow of this experiment is explained. First, the ground is mixed and flattened. Wheel part is set on the ground. Next, wheel part is penetrated in soil tank. Speed of moving wheel part is 0.55mm/s. 2 patterns of the experiment are carried out. One is the pattern that wheel part is penetrated while vibrating. Other is the pattern that wheel part is penetrated without vibration. The force sensor receives value of resistance force from the ground. The number of trials is 10 times. Fig. 5 shows flow of this experiment. Table 1 shows condition of this experiment.

In this experiment, 4 kinds of wheel are used. Fig. 6 shows these wheels. As shown in Fig. 6(c), some protrusions which is called rag are attached to the surface of rag typed wheel. The wheel which is similar rag typed wheel were used by some rovers which carried out the space exploration[5][6]. Therefore, it is important to confirm that the relationship between reg typed wheel and resistance force from the ground. Flange typed wheel has flange on side as shown in Fig. 6(d). Study group of Yamakawa developed the wheel which has flange on side and confirmed that this shape is effective to reduce subsidence amount to the ground from experimental result[7]. In this experiment, flange typed wheel is used in order to confirm that this shape is effective to reduce subsidence amount when vibrating or not. Moreover, two flat typed wheel whose diameters are different are used as shown in Fig. 6(a) and Fig. 6(b). Diameters of flat typed wheel(Large) is same as outer diameters of rag typed wheel and flange typed wheel. Diameters of flat typed wheel(Small) is same as inter diameters of rag typed wheel and flange typed wheel. Experimental results of these wheels are compared with other wheels.

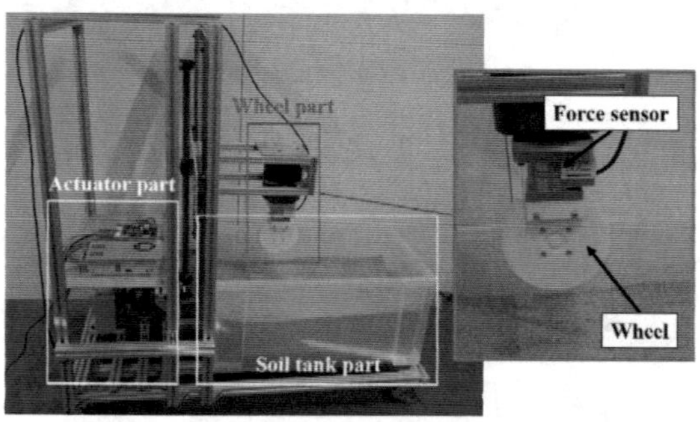

Figure 3. Overview of penetration testing machine.

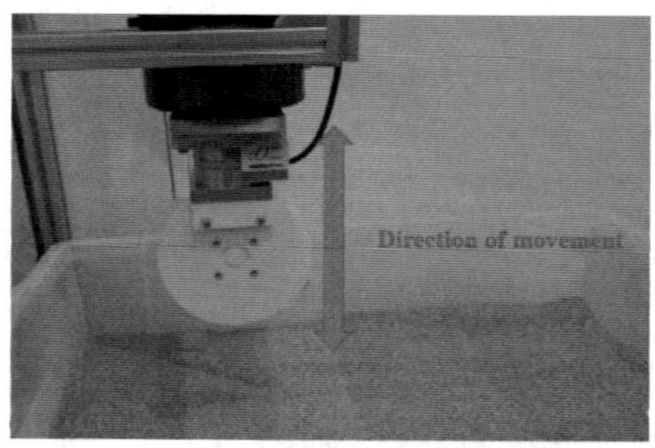

Figure 4. Movement of penetration testing machine.

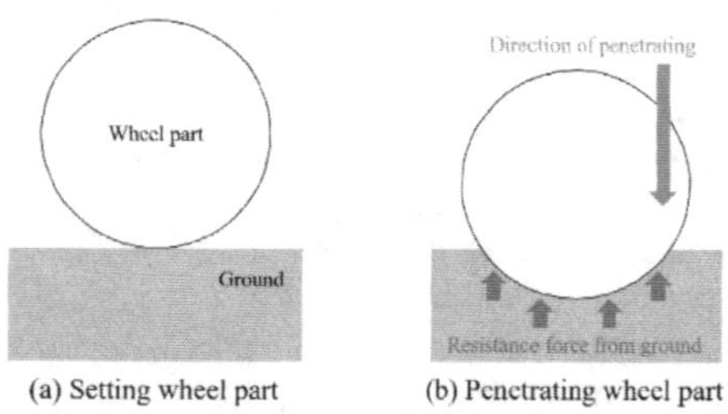

Figure 5. Flow of penetration experiment.

Table1. Condition for experiment

Item	Conditions (value)
Number of trial	10 trials
Penetration speed	0.55 mm/s
Pattern	With vibration Without vibration
Kinds of sand	Silica No.5

IOP Conf. Series: Materials Science and Engineering **904** (2020) 012007 doi:10.1088/1757-899X/904/1/012007

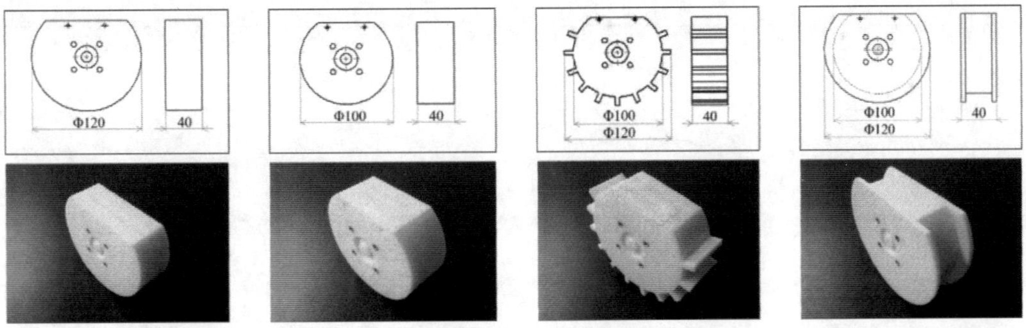

(a) Flat typed wheel (Large) (b) Flat typed wheel (Small) (c) Rag typed wheel (d) Flange typed wheel

Figure 6. Kinds of wheel.

3.3. Results and Discussion

Fig. 7 to Fig. 10 show the graph about distance of sinkage vs resistance force from the ground each shape of wheel. These graphs show data of median value because these trials in each shape of wheel vary hardly in value. In the case of experiment using flat typed wheel(Large) and rag typed wheel, resistance force from the ground which is measured while vibrating is smaller than that which is measured without vibration. Therefore, these shapes of wheel increase subsidence amount to the ground when these wheels vibrate. On the other hand, resistance force from the ground which is measured while vibrating is not smaller than that which is measured without vibration in the case of experiment using flat typed wheel(Small) and flange typed wheel. Therefore, these shapes of wheel hardly sink to the ground when these wheels vibrate.

Fig. 11 and Fig. 12 are the graph which shows results about same outer diameter of wheels. Fig. 13 shows start line of penetration in each wheel. As the result without vibration, flat typed wheel was most difficult to sink to the ground. Rag typed wheel was second most difficult to sink to the ground. Moreover, flange typed wheel was most easy to sink to the ground. The reason is considered that surface of flat typed wheel which touches the ground is the largest in each wheel. As the result with vibration, rag typed wheel was most easy to sink to the ground. Flat typed wheel was most difficult to sink to the ground in the first half of the graph. However, difference about value of resistance force between flat typed wheel and flange typed wheel is small in the second half of the graph.

Fig. 14 and Fig. 15 are the graph which shows results about same inside diameter of wheels. Fig. 16 shows start line of penetration in each wheel. As the result without vibration, rag typed wheel was most difficult to sink to the ground. Flange typed wheel was second most difficult to sink to the ground. Moreover, Flat typed wheel was most easy to sink to the ground. As the result with vibration, flange typed wheel was most difficult to sink to the ground. Therefore, it is considered that flange typed wheel reduces subsidence amount to the ground when vibrating. The reason is considered that flowing sand is blocked by flange.

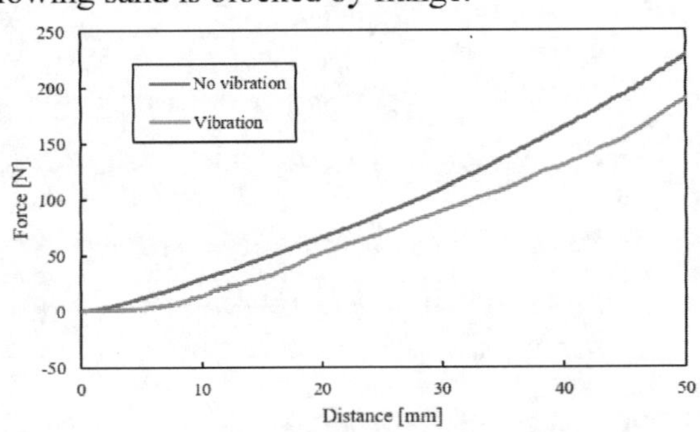

Figure 7. Resistance force vs penetration distance using flat wheel (Large).

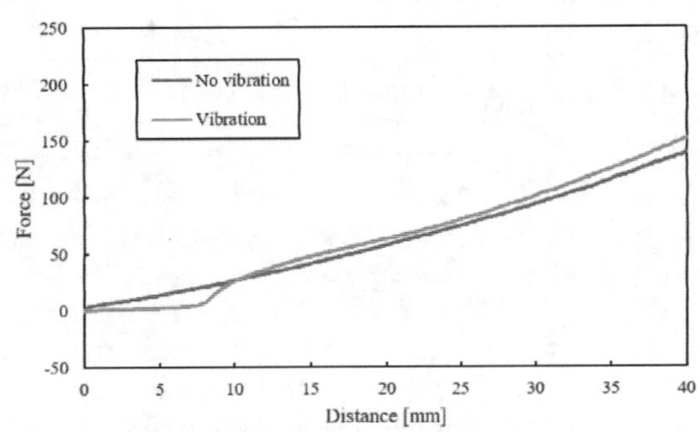

Figure 8. Resistance force vs penetration distance using flat wheel (Small).

IOP Conf. Series: Materials Science and Engineering **904** (2020) 012007 doi:10.1088/1757-899X/904/1/012007

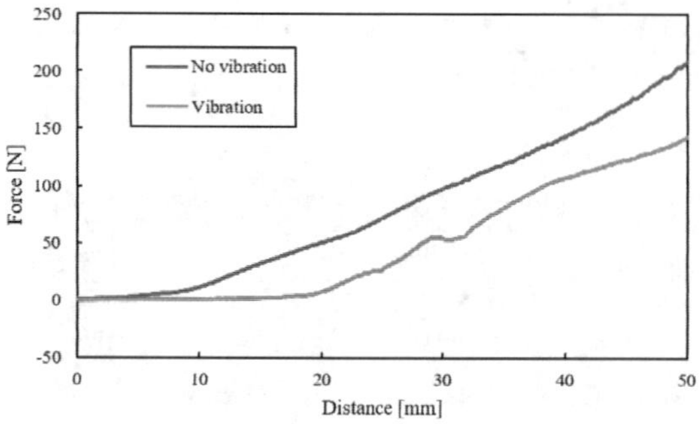

Figure 9. Resistance force vs penetration distance using rag wheel.

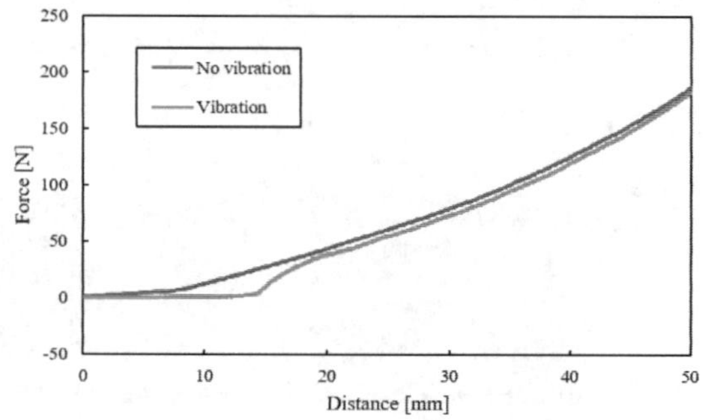

Figure 10. Resistance force vs penetration distance using flange wheel.

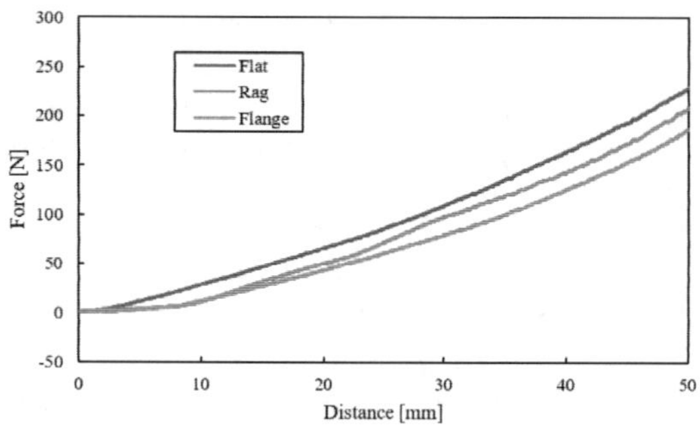

Figure 11. Comparison of resistance force using wheels which are same outer diameter without vibration.

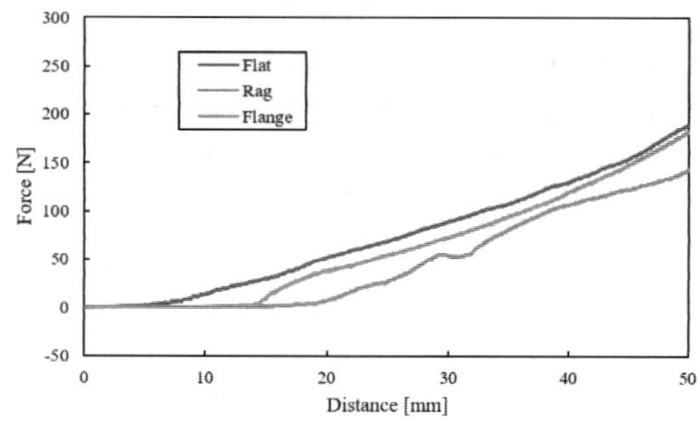

Figure 12. Comparison of resistance force using wheels which are same outer diameter with vibration.

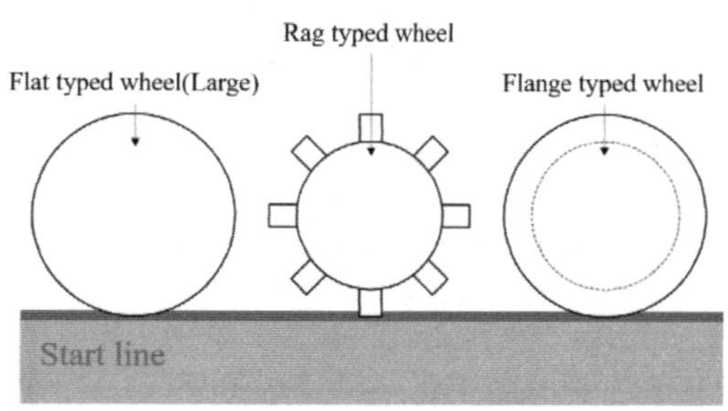

Figure 13. Start line of penetration in each wheel which is same outer diameter.

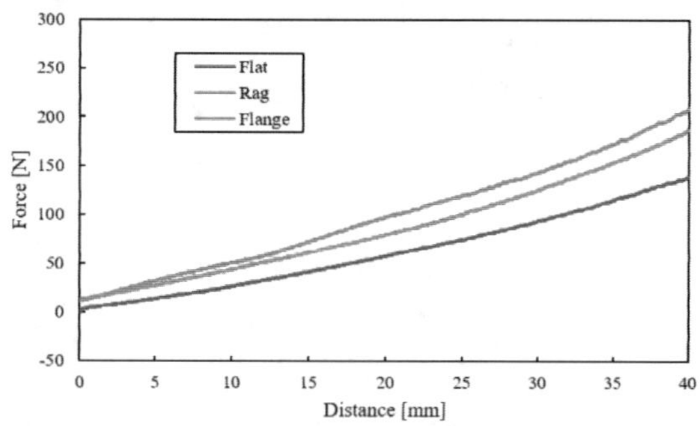

Figure 14. Comparison of resistance force using wheels which are same inside diameter without vibration.

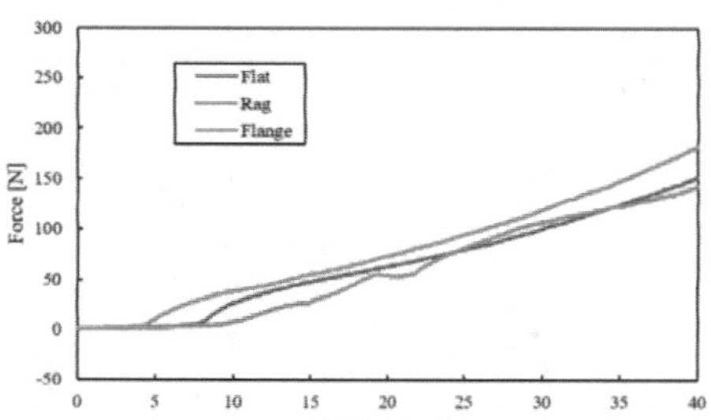

Figure 15. Comparison of resistance force using wheels which are same inside diameter with vibration.

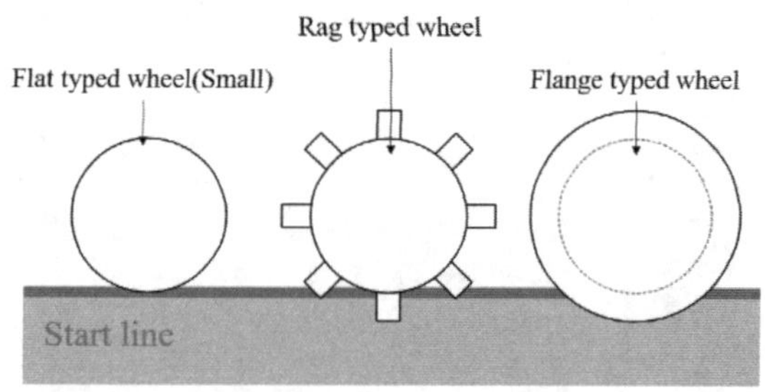

Figure 16. Start line of penetration in each wheel which is same inside diameter.

4. Conclusion

In this study, the relationship between the shape of wheel and resistance force from the ground was confirmed for proposing the wheel which reduce subsidence amount to the ground when vibration is given to the ground. The penetration experiment for measuring resistance force from the ground was conducted. As a result, resistance force from the ground which was measured while vibrating was smaller than that which was measured without vibration in each shape of wheel. The reason is considered that ground becomes soft because sand is flowing by vibration. However, difference of resistance force about flange typed wheel between one with vibration and one without vibration was small. Therefore, the wheel which has flange on side reduced subsidence amount to the ground. The reason is considered that flowing sand is blocked by flange.

From theses experimental results, we could get knowledge about the shape of wheel which reduces subsidence amount to the ground. In the further study, the wheel which reduce subsidence amount to the ground when vibration is given to the ground will be proposed. Moreover, effect of proposed wheel will be confirmed using small wheel typed testbed. In this experiment, the testbed which is attached to proposed wheels will run on the loose ground with slope. Furthermore, movement performance of the testbed will be confirmed.

Acknowledgments
This work was supported by JSPS Grants-in-Aid for Scientific Research JP18K04027.

References

[1] HAKUTO, available from <https://ispace-inc.com/team-hakuto>, (accessed on 2 April, 2020) (in Japanese).
[2] HAYABUSA2 PROJECT, The small exploration rovers, MINERVAII1 (2018), available from <http://www.hayabusa2.jaxa.jp/topics/20180919>, (accessed on 2 April, 2020) (in Japanese).
[3] Yoshimitsu, T., Haruyama, J., Sawai, S., Tsunoda, H., Kuratomi, T., Watanabe, K., Proposal on micro rover system exploring for lunar hole, Proceedings of 11th Space Science Symposium (2011) (in Japanese).
[4] T. Watanabe and K. Iizuka, Study on legged typed rovers with function of vibration/stop to travel on loose soil, presented at 15th ISTVS European-African Regional Conference, September 2019.
[5] NASA, JPL, California institute of technology pasadena, Mars pathfinder, NASA Facts, available from <https://www.jpl.nasa.gov/news/fact_sheets/mpf.pdf>, (accessed on 2 April, 2020) (in Japanese).
[6] Kubota, T., Mobile explorer robot (rover) for lunar or planetary exploration, Journal of JSME, Vol. 104, No. 987 (2001), pp.71-–74 (in Japanese).
[7] J. Yamakawa and M. Akahane, Experimental Research on Wheel Profile to Reduce Sinkage on Sandy Soil, Journal of JSME, Vol. 79, No. 807 (2013), pp.4187-4197 (in Japanese).

IOP Conf. Series: Materials Science and Engineering **904** (2020) 012008 doi:10.1088/1757-899X/904/1/012008

Axial compressor optimization method

Grigorii Popov, Oleg Baturin, Evgenii Goriachkin, Vasilii Zubanov and Andrei Volkov

Samara National Research University, Samara, 443086, Russia

grishatty@gmail.com

Abstract. In the presented work, an algorithm was developed for finding the optimal configuration of the blades of multi-stage axial compressors using 3D CFD modeling as the main tool and using commercial optimization programs. When developing the algorithm, special attention was paid to the development of a method for parameterizing the shape of the blade and a program based on it, which allow you to automatically change the shape of the blade of the axial compressor. They were used by the authors during optimization as a tool that converts variable parameters into the "new" blade geometry. The created algorithm can be used to improve the basic parameters of the compressor (efficiency, pressure ratio, mass flow rate, etc.) due to the correction of the shape of the blade profiles and their position relative to each other. In this case, the algorithm considers the possible presence of various constraints.

Nomenclature

GTE	=	gas turbine engine
MAC	=	multistage axial compressor
FV	=	finite volume
RB	=	rotor blade
RW	=	rotor wheel
GV	=	guide vane
π_c^*	=	pressure ratio
η_c^*	=	compressor efficiency
ΔKs	=	stability margins
n	=	rotational speed
G	=	mass flow rate of the working fluid
α	=	flow angle in absolute motion

1. Introduction

One of the most important tasks in gas turbine engine building is to increase the energy efficiency of the engines (to reduce fuel consumption). It is determined by many multidirectional influencing parameters characterizing all components and the entire engine. One of the most significant is the compressor efficiency. The shortfall in the efficiency of each compressor by 1.0% can increase the specific fuel consumption of a gas turbine engine by up to 0.5% [1].

The compressor efficiency is laid at the stage of its design. Therefore, increasing the efficiency of the design process and the tools used in this process can significantly help to achieve high compressor efficiency.

Recently, compressor design methods have evolved significantly due to the numerical simulations based on the Navier-Stokes equations and the exponential growth of computer capabilities. These factors made it possible to conduct not only a detailed 3D simulation of the working processes of multistage axial compressors, but also to optimize the spatial shape of the blades using mathematical optimization methods.

The latter circumstance allows to create the design variants of the compressor in an automated mode that best satisfy all the design requirements, considering the existing constraints of various kinds. At the same time, the blade that best suits the task can have complex spatial shape (offsets of the sections in axial and circumferential directions, the complex law of changing the chords along the blade height etc.).

There may be an opinion that optimization is a magic tool that allows the engineer to get the best compressor variant for each specific case, with minimal user involvement, which reduces the requirement for the qualification of the designer and the likelihood of errors. However, the use of mathematical optimization in the design and development of the MACs has not yet become widespread. Today, only a small number of successful applications of this method are known, and most of them relate only to individual blade rows and stages.

Among the main difficulties hindering the widespread use of optimization are the following. Firstly, for successful optimization, a 3D numerical model of the MAC workflow is required, which can adequately predict the effect of changes in the design parameters of the compressor on its workflow. Secondly, the mathematical model of the compressor must have an acceptable calculation time, since often at least 1000-3000 iterations with the computational model are necessary for solving the optimization problem. Thirdly, stable parametric models of compressor blades are required coupled with the geometry of the computational domain, which can automatically rebuild the 3D model of the blades. To describe the three-dimensional shape of a compressor blade in a parametric form, many variables are required, and their maximum available number is limited when solving the optimization problem. All the components of the optimization system must be combined within a single automated software that works stably in the entire possible variation range of the parameters with minimal user involvement.

The contradictory nature of the above requirements and the absence of stable working parameterization tools are the main reasons why multicriteria optimization techniques are currently used to a limited extent when modernizing multistage compressors.

In the presented work, the authors set as the goal the development and practical testing of methods and tools for multicriteria optimization of the MAC flowpath elements using numerical parametric three-dimensional models of their working process. Hereinafter, the MAC efficiency criteria are understood as: efficiency η_c^*, pressure ratio π_c^* and gas-dynamic stability margin ΔKs.

2. Review of the Optimization Methods and Tools for the Compressor Working Process

At the first stage, an analysis of the MAC optimization experience was carried out based on the available scientific and technical publications. Summary of the information showed that the usage of optimization methods is one of the most promising ways for improving the design methods and gas-dynamic modernization of the MAC. During the review, the most common optimization methods were found that are used to search for the shape of turbomachinery blades and their relative position in multistage axial compressors: genetic algorithms, conjugate gradient method, artificial neural networks, particle swarm method, indirect optimization method based on self-organization, on which in particular, the IOSO software package is based [2-6].

Summary of the analysis results showed that today there is no unequivocal opinion on which method is best suited for optimizing the MACs. The choice of the method is often based on the traditions of the research team and affordable software. Moreover, the following general elements can be distinguished in all optimization algorithms found during the review:

- a parameterization component used to create a geometric model of the blades and blade rows based on a set of variable parameters;
- a component intended for the automated creation of a FV model of the MAC flowpath with a modified geometry;
- component of the mathematical model of the MAC workflow, which is used to determine the parameters of interest for the investigated object.

In all studies, it is noted that for successful optimization, all these components must be combined into a single computational algorithm.

It was also noted that only one or two blade rows are considered in most works on the optimization of axial compressors, which indicates the difficulties encountered in the application of optimization methods for the entire flowpath of the MAC and confirms the relevance of the work.

3. Parametrization of Compressor Blade Profiles

One of the key technologies of the multicriteria optimization method for the MAC flowpath elements is parameterization of the blade shape. Hereinafter, parameterization refers to a set of independent variables that fully describe the shape of the blade profile of in conjunction with the algorithm for its construction.

The authors of the article developed and implemented their own method of parameterizing compressor blades. It is carried out in 2 stages (Figure 1, Table 1).

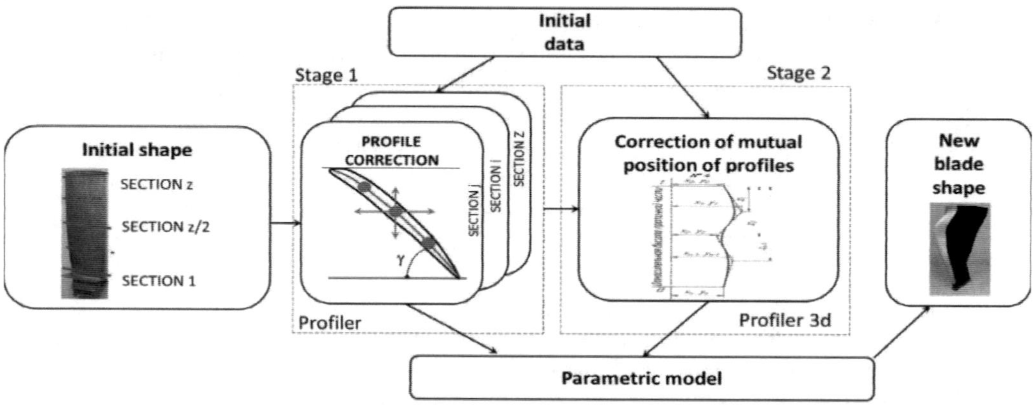

Figure 1. Schematic diagram of the created parameterization scheme.

Table 1. Parameterization stages and their characteristics.

Stage	**1**	**2**
Implemented in the program	Profiler	Profiler 3D
What changes at the stage?	– Stagger angle of the middle line – Position of the control points on the middle line –	– Shift of individual sections in the axial and circumferential direction – Scaling of the profile chord – Scaling of the profile thickness
How is the pattern of shape change set?	The coordinates or parameters of each corrected section and the midline points are directly changed	The dependence of the change of the adjusted parameter along the blade height is set and the parameters of the control points of the dependence are modified

At the first stage, a description of the shape of the planar control sections (cylindrical or conical sections, the rotational axis of which coincides with the rotational axis of the turbomachine) is performed in the *Profiler* program [7]. It was developed at the Department of Aircraft Engine Theory of Samara National Research University [8]. This program can convert the coordinate table, which in the design drawing describes the shape of the blade, into text files of the initial data for constructing design models of turbomachines in the *NUMECA* and *TurboGrid* programs. The *Profiler* also can change the shape of the midline, the stagger angle of the profile, and several other important characteristics of the shape of the blade profile [9].

The second stage of parameterization is carried out in the *Profiler 3D* program developed by the authors of the article [10]. It can perform the followings:

– change the relative position (shifts in axial, radial and circumferential directions) of the control sections of the blade based on the selected stacking law of the sections along the blade height;
– scale the thickness and chords of the cross sections of the blade by correcting the laws of change in scaling factors along the height of the blade.

The algorithm underlying the *Profiler 3D* was built considering the characteristics of the compressor workflow. It is well known that the flow in the blade row has a complex spatial nature, and depends on the loading of the blade rows, which can be different for different stages [11]. Therefore, the parameterization scheme was chosen based on physical ideas about the flow structure in blade rows, so as to be able to purposefully influence the shape of the blades in the characteristic areas of the blade passages: in the flow core, near the endwall regions of hub and shroud.

The second feature is that the number of blade rows in the MAC can be large and exceed the value of 30. Considering that the number of variable parameters during the optimization process is usually limited, the selected parameterization scheme for the MAC blades provides the user with the opportunity to change the number of variable parameters per one row depending on the dimension of the task.

The main idea of a parametric description of the relative position of the sections along the blade height, implemented in the Profiler 3D, is to use the dependence of the variable parameter distribution along the radius $x_i=f(r)$ (for example, the displacement of the section along the coordinate axes) (Figure 2). The linkage changes by correcting the value of the variable in the control points of the dependence $x_i=f(r)$.

In the developed program, the user can fundamentally change the law of distribution of parameters along the radius by choosing the number of control points on the dependence $x_i=f(r)$. In principle, the following laws can be implemented (Figure 2):

– linear – by two control points (shroud and hub);
– dependence of degree 2 – by three control points (shroud, hub and intermediate);
– dependence of degree 3 – by four control points (shroud, hub and two intermediate);
– dependence of an arbitrary degree (with an arbitrary number of control points).

The values of variables in arbitrary points of the dependence $x_i=f(r)$ are calculated by the equation of a polynomial of the corresponding degree [12]. The patterns of parameter changes along the radius can be different for different rows of the same MAC.

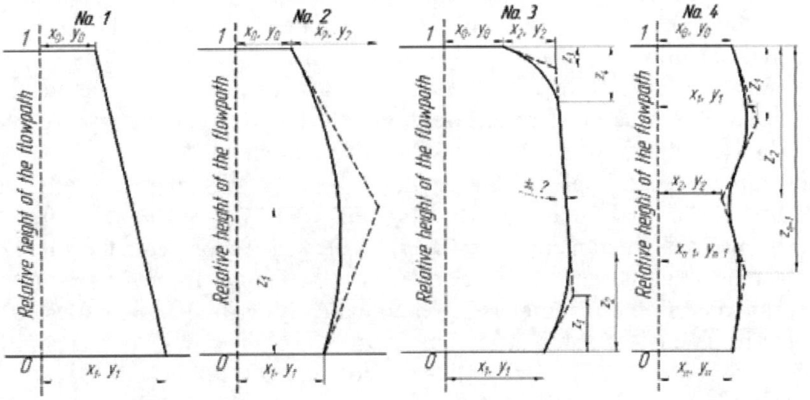

Figure 2. Different ways of describing parameter changes along the radius in Profiler 3D.

The fundamental choice of parameter changing law (the number of control sections) along the height of the blade gives the user an effective tool to reduce the number of variables describing the shape of the profile. As the law becomes more complicated, the user gets more flexibility during optimization, which will help to achieve a better result, but the required number of variables for describing the profile increases significantly. The impact on the radial position of the control points can adjust the degree of deformation along the blade and change the blade to a greater extent where it is really needed (differently in the core and near the endwall regions).

The law of scaling the chords and thicknesses of the blade profiles is set similarly to how the sections are aligned along the blade height.

A coupling of Profiler and Profiler 3D programs (Figure 1) receives the value of independent variables as the input describing the shape of the blades in accordance with the accepted parameterization schemes. As a result, files are generated for each row that describe the geometry of the blades in the *.geomturbo* format adopted in the NUMECA AutoGrid 5 [13]. There, a mesh of finite volumes for the new flowpath is generated. The same program can adjust the axial position of the rows relative to each other, as well as to set, if necessary, a circumferentially variable pitch of the blades. In the process of optimization, the required values of the displacement of the blades in the blade rows are automatically calculated by a special macro and recorded in a script executed during the optimization process.

4. Algorithm for Finding the Optimal Configuration of Multi-Stage Compressor Blades Using a Commercial Optimizer

Based on the literature review and practical experience, an algorithm for multicriteria optimization of the shape and relative position of the MAC blades using commercial optimizer and numerical parametric 3D model of its workflow was developed. The algorithm was implemented based on the Numeca software system [14] and the IOSO optimizer program [15] using in-house programs described above [7, 10].

Any mathematical optimization algorithm or commercial program can be used for the optimization. The developed algorithm is universal and the choice of the best mathematical method for finding the optimum function is not the goal of this study. The authors of the article tested their algorithms using the IOSO program available on the market. Its choice is due to the large number of results of its successful application in the tasks of aircraft engine building, including by the team of authors of the article [6, 16-19]. The specified program was used as a finished commercial product. No upgrades were made to the optimization algorithms. A description of the algorithms used in the program can be found on the website and publications of the program developer [15].

The flow chart of the algorithm is shown in Figure 3. The task of optimizing the MAC is solved iteratively. Before the optimization, variable parameters are set (usually this is the blade geometry using the parameters specified in Section 2), their ranges, criteria (usually the pressure ratio, efficiency, working fluid flow rate or the stability margins) and constraints (strength, structural or technological). The adopted set of variable parameters ensure complete change of the shape of the blade (angles, thicknesses, the shape stacking line of profiles in the radial direction).

The optimization cycle is implemented in the following sequence. Initially, the optimization program (for example, IOSO), based on a list of variable data and a special mathematical algorithm, generates a vector of variable parameters $x = (x_1, \dots, x_n)$, which represents the values of the variables of the blades' parametric model and their relative position for the formed variant of the MAC. The values of the vector of variable parameters are automatically written to text files, which are used as initial data for the programs for parametric construction of blade rows Profiler and Profiler 3D (see section 2). These programs generate files with the geometry of the blades in the *.geomturbo* format, which are then transferred to the NUMECA AutoGrid 5, where the FV grid is created (in accordance with the recommendations of Section 3).

Then, the MAC workflow is calculated for one or more operating modes in the CFD software NUMECA FINE/Turbo. It should be noted that the compressor workflow is calculated using a joint model that considers all stages at the same time and their mutual influence. If the configuration proposed by the optimizer does not work stably for some reason, the CFD solution does not converge at this point and this was considered when the optimizer is working.

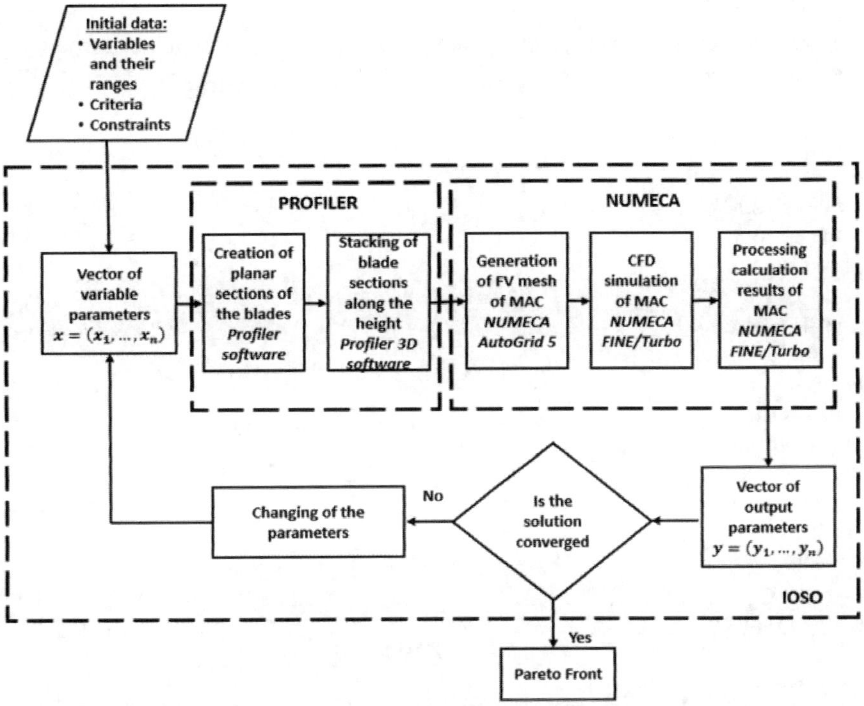

Figure 3. Flow chart of the developed optimization algorithm.

At the end of the calculation, a text file is formed with the values of the relevant compressor operation parameters. The values of these parameters determine the vector of output parameters $y = (y_1, \dots, y_n)$, which is automatically transferred to IOSO. The vector of output parameters may contain the values of the integral parameters of the flow, for example, the flow rate, the values of pressures and temperatures in the cross sections of the MAC flowpath, as well as complex parameters determined by mathematical expressions based on the integral parameters of the flow, for example, the margin of gas-dynamic stability, the efficiency of the MAC. In IOSO, an analysis of the obtained compressor variant and its saving in the search history of the problem are performed based on a distinctive mathematical algorithm. As a result of the analysis, a front of Pareto-compromise solutions is formed (particular solutions in which an improvement of one of the criteria cannot be achieved without deterioration of the others) among the best MAC variants that satisfy the given constraints, and also a vector of variable parameters for a new iteration is created.

The developed optimization algorithm has been repeatedly tested by the authors in solving various industrial problems. A brief description of some tasks, as well as links to publications that describe them in detail, are given in Table 2.

5. Conclusion

The paper describes an algorithm developed by the authors for finding the optimal configuration of the blades of multistage axial compressors using 3D CFD modeling as the main tool and using commercial optimization programs. An important element of the algorithm is the original way of parameterizing the shape of the blades and the program based on it, which allow to automatically change the shape of the axial compressor blades. They were used as an important part of the optimization algorithm as a tool that converts variable parameters into the "new" blade geometry. The created algorithm for searching for the optimal shape of the compressor blades can be used to improve the basic parameters of its operation (efficiency, pressure ratio, flow rate, etc.) by correcting the shape of the profiles of the blades and their position relative to each other. In this case, the method takes into account the possible presence of various constraints.

Table 2. The results of some optimization tasks solved using the developed algorithm.

Example	**1 (HPC)**		**2 (LPC)**	**3 (Three-spool compressor)**
Number of optimized parameters	1	2	2	Number of optimized parameters
Number of constraints	no	4	3	Number of constraints
Number of variable parameters	3	15	61	Number of variable parameters
		Variable parameters		
Stagger angle of the entire blade	+	+	-	Stagger angle of the entire blade
Stagger angles of individual profiles	-	-	+	Stagger angles of individual profiles
Position of the points of the profile middle line	-	-	+	Position of the points of the profile middle line
Shift of individual section relative to the initial position	-	-	+	Shift of individual section relative to the initial position
Chord scaling	-	-	-	Chord scaling
Blade thickness scaling	-	-	-	Blade thickness scaling
		Results		
Efficiency increase	0.3% at 100% mode	1.2% at 95% mode and 0.5 at 100% mode	1.3 at 100% mode	Efficiency increase
Change in the pressure ratio	-	no	+4% (according to the task)	Change in the pressure ratio
Change in stability margins	no	no	no	Change in stability margins
Change in mass flow rate	-3%	no	- 8 % (according to the task)	Change in mass flow rate
Reference	[20]		[21]	[19]

The developed optimization algorithm can solve a problem with an unlimited number of variable parameters, constraints and optimization criteria. However, it is obvious that their increase will significantly complicate the search for the optimum and increase (up to the unacceptable) time to search for a solution. However, in practice, the authors did not solve problems in which the number of optimization criteria was more than two (due to the complexity of analyzing the results with a larger number of variables), and the number of variable parameters was not more than 99 (due to limitations of the available license for the IOSO optimizer program).

Acknowledgments

This work was financially supported by the Ministry of Science and Higher Education within the framework of the state assignment (project number FSSS-2020-0015, "Study of stable and unstable dynamic and vibroacoustic processes in hydraulic and gas systems based on physical and mathematical modeling").

References

[1] Inozemcev, A.A., Nihamkin, M.A. and Sandrackii, V.L. *Osnovy konstruirovanija aviacionnyh dvigatelej i jenergeticheskih ustanovok (Fundamentals of designing aircraft engines and power plants).* Mashinostroenie, Moscow (2008).

[2] Kang, Young-Seok, Park, Tae-Choon, Yang, Soo-Seok, Lee, Sae-Il, Lee and Dong-Ho. "Multi Disciplinary Design Optimization and Performance Evaluation of a Single-Stage Transonic Axial Compressor." *ASME Turbo Expo 2012: Turbine Technical Conference and Exposition* GT2012-69252 (2012): pp. 361-369. DOI: 10.1115/GT2012-69252.

[3] Zhang, Pengfei, Lu, Juan, Wang, Zhiduo, Song, Liming and Feng, Zhenping. "Adjoint-Based Optimization Method With Linearized SST Turbulence Model and a Frozen Gamma-Theta Transition Model Approach for Turbomachinery Design." *ASME Turbo Expo 2015: Turbomachinery Technical Conference and Exposition* GT2015-42582 (2015): V02BT39A019. DOI: 10.1115/GT2015-42582.

[4] Astrua, Pio, Piola, Stefano, Silingardi, Andrea and Bonzani, Federico. "Multi-Objective Constrained Aero-Mechanical Optimization of an Axial Compressor Transonic Blade." *ASME Turbo Expo 2012: Turbine Technical Conference and Exposition* GT2012-68993 (2012): pp. 241-252. DOI: 10.1115/GT2012-68993.

[5] Sonoda, Toyotaka, Endicott, Giles, Arima, Toshiyuki and Olhofer, Markus. "A Novel Transonic Fan Swept Outlet Guide Vane Using 3D Design Optimization." *ASME Turbo Expo 2014: Turbine Technical Conference and Exposition* GT2014-25857 (2014): V02AT37A022. DOI: 10.1115/GT2014-25857.

[6] Egorov I.N., Kretinin G.V., Fedechkin K.S. "Multi-Level Robust Design Optimization Fan", *Workshop CEAS*, Vrije Universiteit Brussels (VUB), Brussels, Belgium, October 28-29, 2010.

[7] Shabliy, L.S. "Software tool for creating and modifying computer models of turbomachine blades Profiler". Certificate of state registration of a computer program No. 2013617453. 2013.

[8] Samara University. Access mode: https://ssau.ru/english/.

[9] Dmitrieva, I.B. and Shabliy, L.S. "Programma PROFILER dlya sozdania i modifikacii lopatok turbomoshin v ramkah zadach optimizacii." *Problemy i perspectivy razvitia dvigatelestroenia: materialy dokladov megdunarodnoy nauchno-tehnicheskoy konferencii* (2014): pp. 201 – 203.

[10] Popov, G.M., Gorjachkin, Baturin, O.V. Novikova, Ju.D. and Volkov, A.A. "Program for parametric change of 3D shape of axial turbomachine blades Profiler 3D". Certificate of state registration of a computer program No.2017613128. 2017.

[11] Tihonov, N. T., Musatkin, N. F. and Matveev, V. N. *Teorija lopatochnyh mashin aviacionnyh gazoturbinnyh dvigatelej (Theory of blade machines of aircraft gas turbine engines).* Samarskij gosudarstvennyj ajerokosmicheskij universitet im. S.P. Koroleva, Samara (2001).

[12] Flanders, Harley. Elementary Functions and Analytic Geometry. Elsevier (1973).

[13] Numeca AutoGrid, Access mode: https://www.numeca.com/product/autogrid5

[14] Numeca FineTurbo, Access mode: https://www.numeca.com/product/fineturbo

[15] IOSO Optimization Technology. Access mode: http://www.iosotech.com.

[16] Popov, G., Matveev, V., Baturin, O., Novikova, Y., Kolmakova, D. and Zubanov, V. "Estimation of the Effect of 3D Grid Parameters on the Simulation of the Working Process of Axial Turbines." In *Proceedings of 8th International Conference on Simulation and Modeling Methodologies, Technologies and Applications* Vol. 1(2018): pp. 371-377. DOI: 10.5220/0006890203710377.

[17] Popov, Grigorii, Goriachkin, Evgenii, Kolmakova, Daria and Novikova, Yulia. "Multicriteria Optimization of Axial Low Pressure Compressor of Gas Turbine Power Plant." *ASME Turbo Expo 2016: Turbine Technical Conference and Exposition* GT2016-57856 (2016):

V02CT45A028. DOI10.1115/GT2016-57856.

[18] Popov, G.M., Egorov, I.N., Dmitriev, D.N., Goriachkin, E.S. and Volkov, A.A. “Complex gas dynamic optimization of a three spool axial compressor of an industrial gas turbine engine”. In *Proceedings of SME Gas Turbine India Conference*. GTIndia2019-2394.

[19] Baturin, Oleg, Tkachenko, Andrei, Krupenich, Ilia, Popov, Grigorii and Goryachkin, Eugene. "Identifying the Approach to Significantly Improve the Performance of NK-36ST Gas Turbine Power Plant." *ASME Turbo Expo 2017: Turbine Technical Conference and Exposition* GT2017-64836 (2017): V003T23A010. DOI: 10.1115/GT2017-64836.

[20] Marchukov, E., Egorov, I., Popov, G., Baturin, O., Goriachkin, E., Novikova, Y. and Kolmakova, D. "Improving of the working process of axial compressors of gas turbine engines by using an optimization method", *IOP Conference Series: Materials Science and Engineering*. Vol. 232. No.1 (2017): Article number 012041. DOI 10.1088/1757-899X/232/1/012041

[21] Goryachkin, E., Popov, G., Baturin, O. and Kolmakova, D. "Three-stage low pressure compressor modernization by means of optimization methods." *ASME Turbo Expo 2015: Turbine Technical Conference and Exposition* GT2015-43384 (2015): V003T23A010. DOI: 10.1115/gt2015-43384.

IOP Conf. Series: Materials Science and Engineering **904** (2020) 012009 doi:10.1088/1757-899X/904/1/012009

The numerical simulation of the changes of lock-in phenomenon produced from an in-line oscillating circular cylinder by the vortex method

Yoshifumi Yokoi

National Defense Academy of Japan
1-10-20 Hashirimizu, Yokosuka, Kanagawa, Japan 239-8686

yokoi@nda.ac.jp

Abstract. This study carries out a numerical simulation about changes of the lock-in phenomenon of the single circular cylinder which is oscillating in the direction of a flow. As for the calculation method of this study, the vortex method was used. The calculation was carried out by Reynolds number $Re = 500$. The oscillating amplitude ratio was $2a/d = 0.25$ and the range of oscillating frequency ratio f/f_K was 0 to 3. In order to know the reliability of the calculation method, the flow aspect and fluid force of the stationary single circular cylinder were investigated, and it compared with the existing experimental result. The result was well in agreement. Also in calculation of the oscillating single circular cylinder, two typical flow patterns when a lock-in occurs were expressed. The relationship of the drag and lift was also found. The aspects of the flow of the situation which the lock-in does not produce were investigated. In the situation which the lock-in does not produce, there is no particular flow pattern and it became clear that two typical vortex shedding configurations interchanged and appeared. It was found that the character of such a complicated flow can also be expressed by the vortex method.

1. Introduction

If a circular cylinder is placed into a steady flow without a time change, vortices will be discharged alternately. And a Karman vortex street is formed behind the circular cylinder. The Karman vortex is a very stable vortex street, and the vortex corresponding to the flow velocity is formed. That can also be said to be a natural synchronous phenomenon. Many of flows which exist really are what are called unsteady flow to which speed is changed in time. When unsteady, it is thinkable that the characteristics of the phenomenon differ compared with the case of being steady. It is industrially important to grasp the fluid force characteristic and the vortex shedding characteristic of the object put on the unsteady flow field. However, in order to realize an unsteady flow with sufficient accuracy, serious troubles are required in a laboratory. In order to experiment simple, in the laboratory, the object which exercises in the direction of flow was installed into the steady flow, so the relative unsteady flow is made. If the circular cylinder which is oscillating in the flow is placed, the vortex shedding which synchronized with circular cylinder oscillating frequency will be observed. This phenomenon was called "lock-in phenomenon" and, as for this "interference of the flow velocity change", research has been done by many researchers [1 - 4]. It is known that the flow pattern in the lock in state is divided roughly into the "alternate vortex shedding type" and the "simultaneous vortex shedding type". Although there are many studies treating the aspect of the flow of the state of the lock-in, there are few studies treating the aspect of the flow of a non-lock-in state. The flow of the non-lock-in state is complicated. So, it is important to observe such flows. Probably, the special flow pattern of the non-lock-in state does not exist, but is considered that the alternate vortex shedding and the simultaneous vortex shedding have repeated the appearance intermittently. By flow visualization experiment, Yokoi and Hirao [5] are observing intermittent replacement of the alternate vortex shedding and the simultaneous vortex shedding, and have reported the rate. The author carried out the simulation for those lock-in states, and showed the validity of the vortex method. In order to study the phenomenon, the study which investigates fluid force is interesting, and important. Although the experimental study is difficult therefore, the numerical simulation using a computer is expected. In this study, the result of having carried out the numerical simulation about the flow aspect and fluid force in the non-lock-in state is reported.

2. Numerical calculation

The numerical experiment apparatus was consisted of simulation software and a notebook type computer (NEC; LaVie LC958/T) as calculation hardware which are on the market. The software which named 'UzuCrise 2D ver.1.1.3 rev.H (College Master Hands Inc., 2006)' is used. This software used the vortex method which is based on the Lagrangian analysis. Since the vortex method is the grid-less method, it is suitable for the unsteady problem of such moving boundary. The vortex method is a direct viscid-inviscid interaction scheme, and the emanation of velocity shear layers due to boundary layer separation is represented by introduction of discrete vortices with viscous core step by step of time. In the present study, the flow was assumed incompressible and two-dimensional flow field. The configuration of circular cylinder was represented 40 vortex panels using a boundary element method. The separating shear layers were represented the discreet vortices, which were introduced at the separation points. The details of calculation technique of vortex method and accuracy of calculation are shown in Kamemoto [6, 7].

In the present study, the calculations were performed at the two-dimensional flow field for incompressible and viscous flow. A cylinder diameter d and main flow velocity U were determined as 16 mm and 0.04 m/s so that it could compare with the previous experimental result [8]. Since water is assumed as for test fluid, Reynolds number Re becomes 500. The configuration of circular cylinder was represented 40 vortex panels using a boundary element method. Its vortex panels are provided equally. Every calculation continued to until non-dimensional time $T = 200$.

The main parameters of numerical experiment were the oscillation amplitude ratio $2a/d$ (a: half-amplitude of cylinder motion, d: cylinder diameter), the oscillation frequency ratio f/f_K (f: cylinder oscillation frequency, f_K: natural Karman vortex shedding frequency). The oscillation amplitude ratio is three kinds, is 0.00 and 0.25, respectively. Here, when the oscillating amplitude ratio is "0.0", it means circular cylinder stationary. The oscillation frequency ratio is 15 kinds, is from 0.2 to 3.0 every 0.2 steps.

The calculation experiment was performed in the following procedures. The case of the stationary circular cylinder is carried out first and the case where the circular cylinder is oscillating in the direction of a flow after that is carried out. In order to grasp the flow field on the computer, and to determine for the vortex shedding frequency from the stationary single circular cylinder, it needs to be calculated in the case of the stationary circular cylinder. Secondly, the oscillating amplitude ratio is defined ($2a/d = 0.25$), the oscillating frequency ratio is varied, and systematic numerical computation is performed. The arrangement of circular cylinders is shown in Fig. 1.

3. Results and discussions

In order to verify the calculation software to be used, the experimental result and calculation result of a stationary circular cylinder were compared. The time history of the fluid force (a drag component and lift component) of acting on a stationary circular cylinder is shown in Fig. 2. Figure 2 (a) is a time history from the non-dimensional time $T = 0$ to $T = 200$, and Fig. 2 (b) expands the section from the non-dimensional time $T = 150$ to $T = 200$. The drag oscillation lurking in periodic oscillation and lift oscillation of the lift produced by formation of a Karman vortex street is expressed. The relationship whose oscillation frequency of the drag is twice the oscillation frequency of the lift is shown. The following results were obtained after non-dimensional time $T = 150$. Here, the diagram which defines the quantity about the fluid force is shown in Fig. 3. The average value of drag coefficient was $C_{DAVE} = 1.08$ and the root mean square value of amplitude of drag coefficient was $A_{CD} = 0.15$. The average value of lift coefficient was $C_{LAVE} = 0.00$ and the route mean square value of amplitude of lift coefficient was $A_{CL} = 0.73$. The average Strouhal number for which it determined from the lift oscillating period T was $St_{AVE} = 0.26$. The Strouhal number is defined by $St = fd/U = d/\Delta tU = 1/T$. When the Reynolds number is $Re = 500$, it is known that the experimental value of Strouhal number St is about 0.2. Although it seems that this calculation result is highly calculated as compared with an experimental result, it seems that this calculation has obtained the comparatively good calculation result since a two dimensional calculation result becomes higher than an experimental result about 30 to 40%. When the Karman vortex shedding frequency f_K was calculated from the average value of Strouhal number, the Karman vortex shedding frequency was $f_K = 0.65$ Hz.

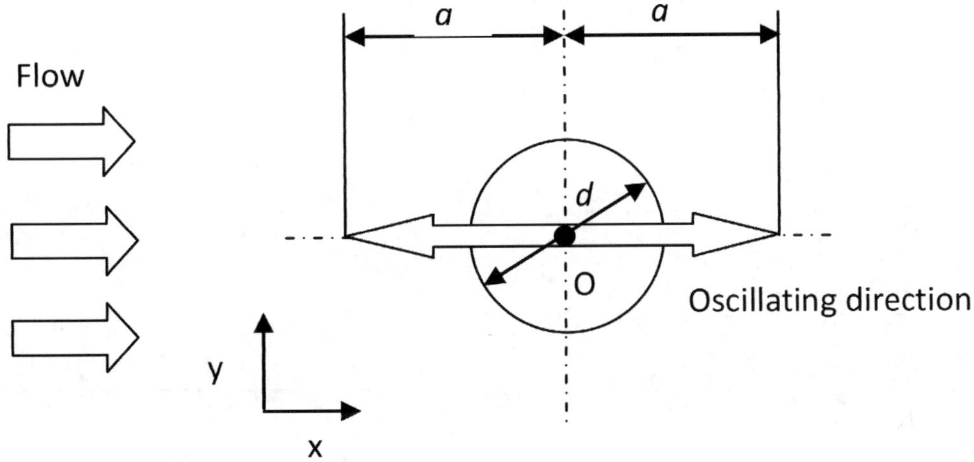

Figure 1. Coordinate system and definition of symbols (O: the coordinate origin).

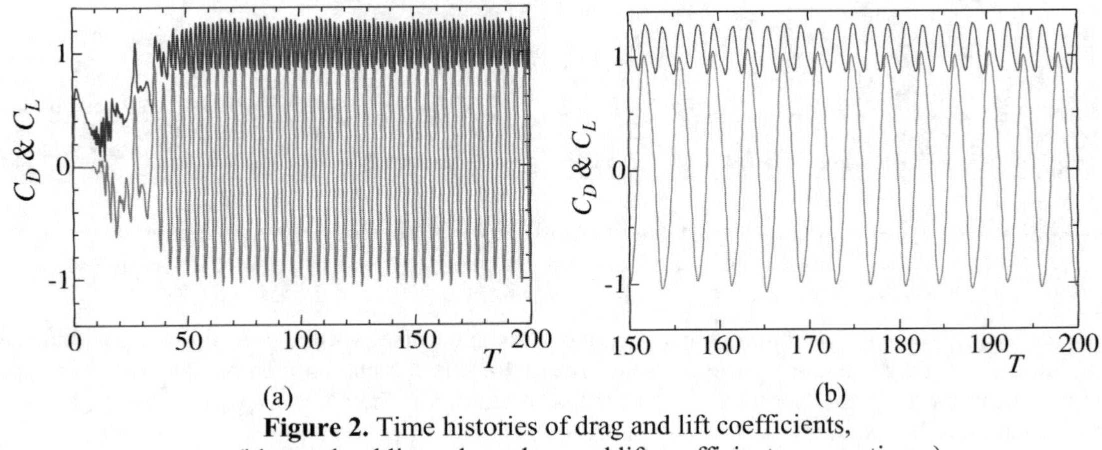

Figure 2. Time histories of drag and lift coefficients,
(blue and red lines show drag and lift coefficients, respectivery).

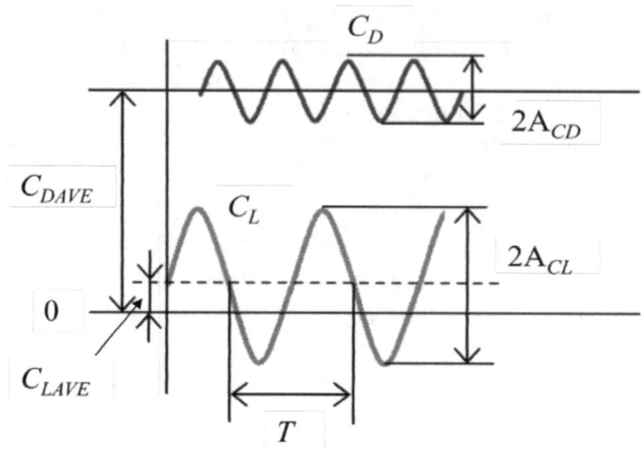

Figure 3. The definition of the magnitude of the drag coefficient or the lift coefficient, and the definition of the oscillating period.

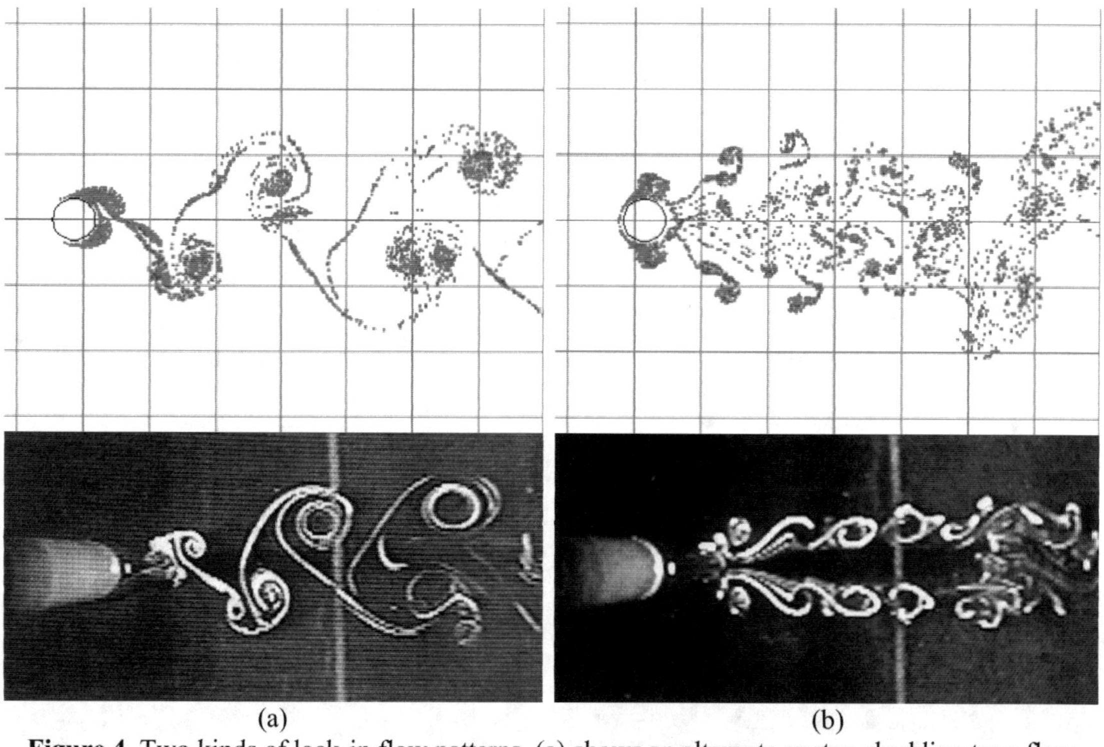

Figure 4. Two kinds of lock-in flow patterns, (a) shows an alternate vortex shedding type flow pattern and (b) shows the simultaneous vortex shedding type flow pattern, respectively.

It calculated in the condition which produces the alternate vortex shedding lock-in and the simultaneous vortex shedding lock-in for authorization of this calculation method. The result is shown in Fig. 4 with a previous experimental result [8]. The feature of the alternate vortex shedding lock-in and the feature of the simultaneous vortex shedding lock-in can be seen well expressed in calculation. The natural Karman vortex lock-in, the alternate vortex shedding lock-in, and the simultaneous vortex shedding lock-in are in the state where the flow pattern is being fixed. In such the state, it was found that the vortex method can be suitable good.

It is observed by the flow visualization experiment [5] that two flow patterns interchange intermittently and appear in the state of the non-lock-in at the time of circular cylinder oscillation. In calculation of the vortex method, it is one of the most interesting matters whether such a state can be expressed. The non-lock-in state is divided into the following three regions. It is the case of circular cylinder oscillation frequency lower than the circular cylinder oscillation frequency which the alternate vortex shedding lock-in produces. The aspect of the flow in this case is an alternate vortex shedding similar to the Karman vortex. It is the case of circular cylinder oscillation frequency higher than the circular cylinder oscillation frequency which the simultaneous vortex shedding lock-in produces. Since the aspect of the flow in this case has strayed off from the simultaneous vortex shedding lock-in, it is an alternate vortex shedding similar to the Karman vortex. In other than these, replacement of those flow patterns is generated. Accordingly, it is the case of circular cylinder oscillation frequency lower than the circular cylinder oscillation frequency which the simultaneous vortex shedding lock-in produces in circular cylinder oscillation frequency higher than the circular cylinder oscillation frequency which the alternate vortex shedding lock-in produces. For imagining the aspect of the flow pattern, it is important to investigate the fluid force.

The time histories of the drag coefficient C_D and the lift coefficient C_L in case oscillating amplitude ratios $a/d = 0.25$ are shown in Fig. 5. As for the time history, the range by the non-dimension time 50 - 100 is shown, a red line shows the lift coefficient, a blue line shows the drag coefficient, and a black line shows the motion of circular cylinder, respectively. It is found with the increase in the oscillation frequency ratio that change is looked at by the magnitude of the amplitude of the drag coefficient and the magnitude of the amplitude of the lift coefficient. When the oscillation frequency ratio is small, it can be seen that the amplitude of lift coefficient is larger than the amplitude of drag coefficient. On the other hand, when the oscillation frequency ratio is large, the amplitude of drag coefficient is seen be larger than the amplitude of lift coefficient. Here, the "lock-in" is that the vortex shedding frequency from the circular cylinder synchronizes with circular cylinder oscillation frequency. So, it takes notice of the relationship between the waveform of circular cylinder oscillation, the waveform of a drag coefficient, and the waveform of a lift coefficient. The case where the lift coefficient is oscillating with the period twice the period of the circular cylinder oscillation is the alternate vortex shedding type lock-in. On the other hand, the case where the drag coefficient is oscillating synchronizing with the period of the circular cylinder oscillation is the simultaneous vortex shedding type lock-in. In the magnitude relationship of the lift coefficient and the drag coefficient, when the amplitude of the lift coefficient is large, it becomes the alternate vortex shedding, and when the amplitude of the drag coefficient is large, it becomes the simultaneous vortex shedding. In Fig. 5, the alternate vortex shedding type lock-in is shown in Fig. 5(g), and the simultaneous vortex shedding type lock-in is shown in Figs. 5(k) - 5(o).

The discovery which is the worthiest by this study was obtained. The example is shown in Fig. 6. It shows the flow pattern at the time of oscillation frequency ratio $f/f_K = 1.6$. Figure 6 (a) is visible to the alternate vortex shedding type of mushroom shape. Figure 6 (b) seems for a twins vortex to incline and to carry out the vortex shedding simultaneously. The time history of the fluid force of this state is shown in Fig. 5 (h). The oscillating period of the drag coefficient synchronizes with the circular cylinder oscillating period. On the other hand, the oscillating period of the lift coefficient is 1.5 times the circular cylinder oscillating period, and has strayed off from the lock-in. As a result, although the vortex shedding near the circular cylinder is carried out simultaneously, the phenomenon in which the discharge direction changes occurs. So, the vortex shedding which occurs simultaneously will change the discharge direction like a pendulum. Since the deflection width of the pendulum depends on the magnitude of the amplitude of the lift coefficient, when the amplitude is small, the direction of a vortex shedding becomes in the direction of a direct down-stream. Accordingly, the vortex to which the boundary layer separation rolled simultaneously is discharged by the balance of fluid force becomes alternate arrangement, or becomes simultaneous arrangement. It was able to express replacement of the vortex shedding in the vortex method.

IOP Conf. Series: Materials Science and Engineering **904** (2020) 012009 doi:10.1088/1757-899X/904/1/012009

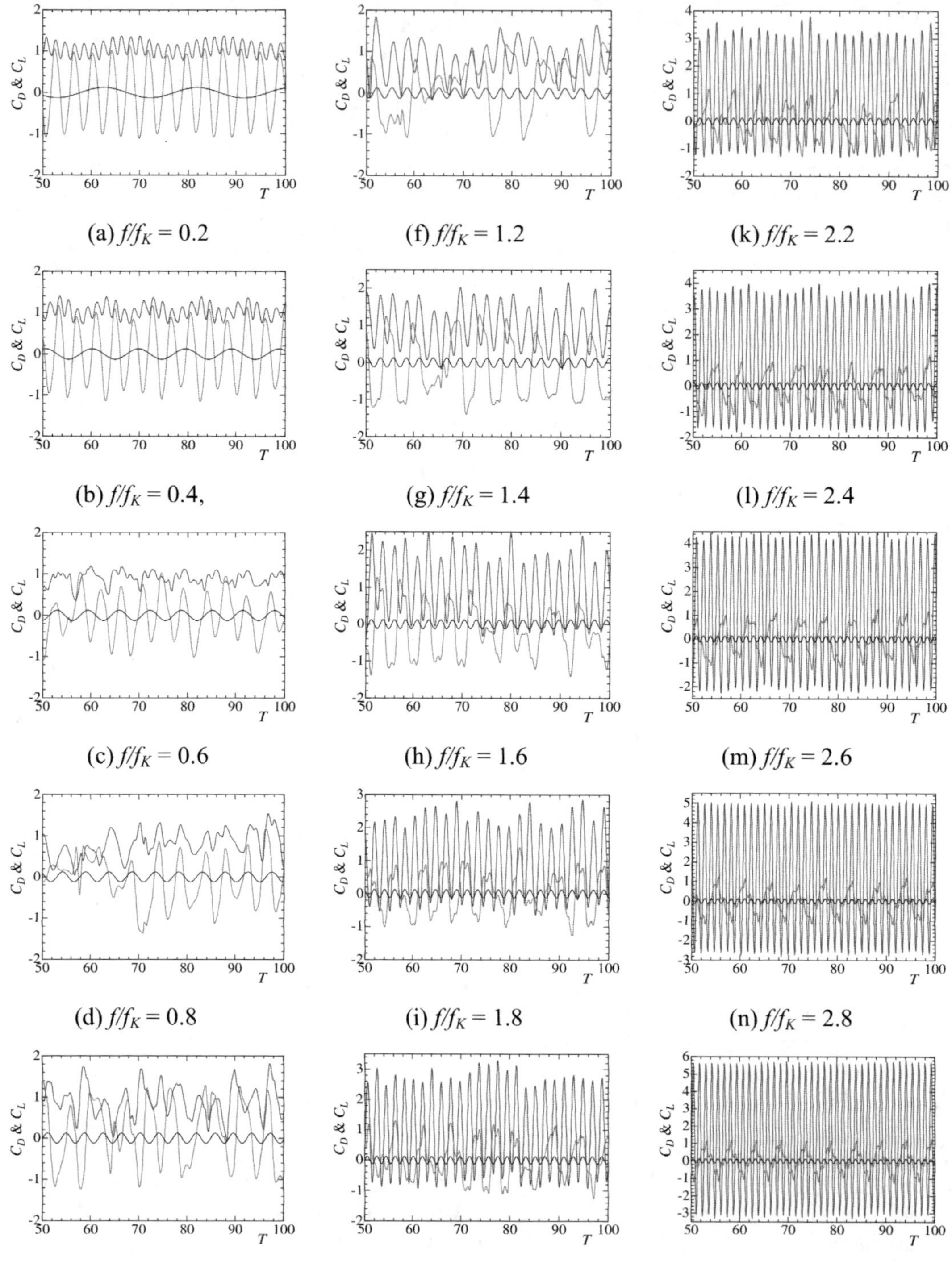

(a) $f/f_K = 0.2$ (f) $f/f_K = 1.2$ (k) $f/f_K = 2.2$

(b) $f/f_K = 0.4$, (g) $f/f_K = 1.4$ (l) $f/f_K = 2.4$

(c) $f/f_K = 0.6$ (h) $f/f_K = 1.6$ (m) $f/f_K = 2.6$

(d) $f/f_K = 0.8$ (i) $f/f_K = 1.8$ (n) $f/f_K = 2.8$

(e) $f/f_K = 1.0$ (j) $f/f_K = 2.0$ (o) $f/f_K = 3.0$

Figure 5. Time histories of drag and lift coefficients in the case of oscillation amplitude ratio a/d = 0.25 (blue and red lines show drag and lift coefficients, black line is cylinder motion).

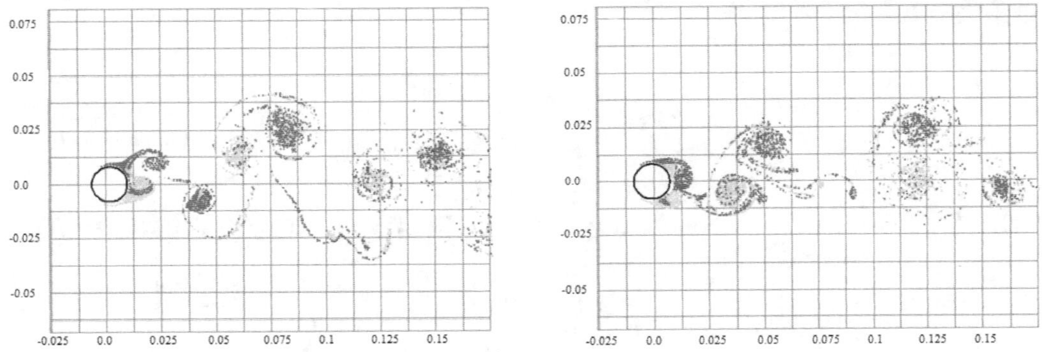

(a) (b)

Figure 6. Instantaneously flow pattern on $f/f_K = 1.6$; (a) non-dimensional time T=56.16, (b) non-dimensional time T= 88.92.

4. Conclusions

A systematic numerical simulation of the flow around a circular cylinder which oscillates in the direction of a flow using the vortex method was performed. The following conclusions were obtained.

(1) The flow pattern of two kinds of lock-in states and the flow pattern of natural lock-in state were well in agreement with the aspect of the previous experiment visualization result.

(2) In the magnitude relationship of the lift coefficient and the drag coefficient, when the amplitude of the lift coefficient is large, it becomes the alternate vortex shedding, and when the amplitude of the drag coefficient is large, it becomes the simultaneous vortex shedding.

(3) The replacement of two flow patterns was expressed by the pendulum vortex shedding phenomenon and the magnitude of the lift which acts on the circular cylinder.

(4) It was found that the vortex method can be suitable also at such a complicated flow field.

(5) As future work, to investigate the dependability of Reynolds number and verification experiments of the numerical computation result are desired.

References

[1] O. M. Griffin and M. S. Hall, "Review-Vortex Shedding Lock-on and Flow Control in Bluff Body Wakes", *Trans. of ASME*, **113**, 526-537 (1991).A reference

[2] R. King, "A Review of Vortex Shedding Research and its Application", *Ocean Engineering*, **4**, 141-171 (1997).

[3] A Okajima, "Flow-Induced Vibration of a Bluff Body", *Trans. of JSME* Series B, **66**-644,948-953 (2000).

[4] A Okajima, "Vortical Flow and Fluiddynamic Characteristics of an Oscillating Bluff Body", T*rans. of JSME* Series B, **66**-644, 948-953 (2000).

[5] Y. Yokoi and K. Hirao, "The appearance of two lock-in states in the vortex flow around an in-line forced oscillating circular cylinder", *EPJ Web of Conferences*, 67, 02131 (2014).

[6] K. Kamemoto, "The expandability of the vortex method as a turbulent flow model (the first part: to think a basic of vortex method)", *Computational Fluid Dynamics*, **2**-1, 20-29 (1993).

[7] K. Kamemoto, "The expandability of the vortex method as a turbulent flow model (the latter part: to lock up the flow by vortex method)", *Computational Fluid Dynamics*, **2**-2, 28-39 (1994).

[8] Y. Yokoi and K. Hirao, "Vortex Flow Around an In-Line Forced Oscillating Circular Cylinder", *Trans. Japnan Society of Mechanical Engineers Series B*, **74**-746, 2099-2108 (2008).

IOP Conf. Series: Materials Science and Engineering **904** (2020) 012010 doi:10.1088/1757-899X/904/1/012010

Wear of a high pressure layered pipe by a rough rigid bush

K E Kazakov[1] and S P Kurdina[2]

[1] Ishlinsky Institute for Problems in Mechanics of the Russian Academy of Sciences, Moscow, Russia
[2] Bauman Moscow State Technical University, Moscow, Russia

E-mail: kazakov-ke@yandex.ru, kurdinasp@bmstu.ru

Abstract. The article describes the solution to the problem of linear wear of a high pressure elastic pipe with an elastic coating by rigid bush worn on the pipe so that there is no gap between them. It is assumed that both the thickness of the coating in undeformed state and the shape of the inner surface of the bush are variable and depend on the coordinate along the axis of pipe. In the case when indicated characteristics are described by complex rapidly changing functions, standard approaches to solving such a problem turn out to be ineffective. The article describes an approach based on the use of special representations and basic functions that take into account the presence of complex characteristics. In the expression for contact pressures, which is the result of solving the problem and allows us to further evaluate the resource and durability of such a connection, the functions that describe the surfaces shapes are highlighted in explicit form. This allows us to perform exact calculations in cases where the forms are described by complex functions.

1. Introduction

In real constructions, pipes loaded with internal pressure are often used. External conditions determine the use of multilayer pipes, in which different layers serve for different purposes. Such pipes can be fastened with bushings, however, during operation due to vibration or torsion of the bush around the pipe, wear occurs, leading to a thinning of the outer layer. To assess the durability of such a connection, it is necessary to monitor the evolution of the stress-strain state of the pipe and evaluate its wear.

2. Mathematical model of contact-wear problem

Consider elastic pipe with inner radius r in and the thickness h in covered by thin elastic layer of variable thickness $h_{\text{out0}}(z)$ and inner radius $r_{\text{out}} = r_{\text{in}} + h_{\text{in}}$. We put rigid bush on such a tube. Inner radius of the bush is variable, depend on the coordinate z and less or equal to outer radius of pipe with coating in undeformed state, i.e. $g(z) \leq r_{\text{out}} + h_{\text{out0}}(z)$. We assume that thickness $h_{\text{out0}}(z)$ is much smaller than bush length $2a$, inner radius of main pipe r_{in}, and thickness of main pipe h_{in}. The coating is considered soft compared to the lower layer (its rigidity not exceed the rigidity of lower layer [1]). It is also assumed that there is a smooth contact between layers and between bush outer layer along z axis. At the time τ_0 bush start to vibrate or rotate with angular velocity ω. Due to friction between the coating and the bush coating begins to wear out. At the same time we begin to load the pipe with internal pressure $P_{\text{in}}(t)$. The scheme for such a problem for the case of rotation is presented in the figure 1. It is obvious that the stress-strain state of the layered pipe and, in particular, the contact

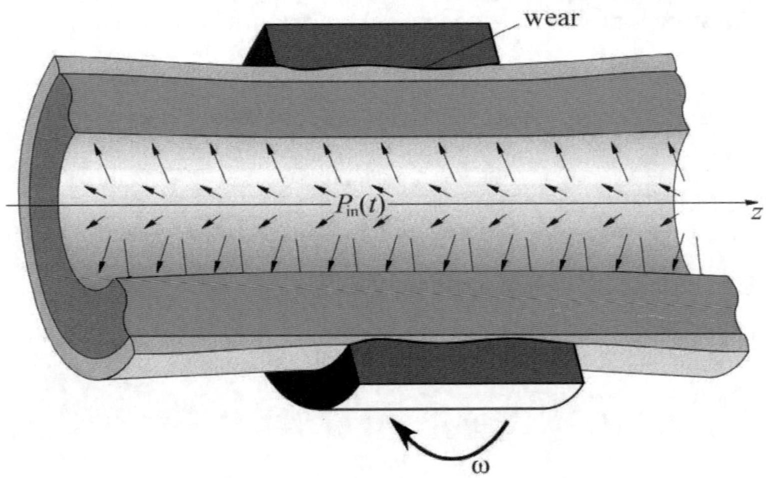

Figure 1. Contact-wear problem.

pressures under the bush will change over time. Further we will assume that the contact-wear region is constant and bounded by the boundaries $-a$ and a (coordinates of left and right boundaries of the bush). We will construct solution only for the case when wear is less then coating thickness.

The movement of the outer boundary of the coating under the bush consists of displacement due to load and displacement due to wear.

A number of experiments show that the wear velocity linearly depends on the relative velocity of the contacting surfaces and on the normal load and is inversely proportional to the hardness of the material [2–6]:

$$\upsilon_{\mathrm{w}}(z,t) = -\frac{k_{\mathrm{w}} V q(z,t)}{H(z)},$$

where $H(z)$ is coating hardness, $q(z,t)$ is contact pressure, k_{w} is experimentally determined constant. The velocity V can be estimated with fretting-wear of surfaces, when the angular velocity is absent, however, the coating wears out due to vibrations of one body relative to another. In the case of rotation $V = \omega(r_{\mathrm{out}} + \bar{h}_{\mathrm{out}}) \approx \omega r_{\mathrm{out}}$, where $\bar{h}_{\mathrm{out}} = h_{\mathrm{out}0}(z_0)$, z_0 is any point in $[-a,a]$ (note that $h_{\mathrm{out}0}(z) \ll r_{\mathrm{out}}$ for all $z \in [-a,a]$ and hence $\bar{h}_{\mathrm{out}} \le \max_{z\in[-a,a]} h_{\mathrm{out}0}(z) \ll r_{\mathrm{out}}$). Then the vertical displacement of outer surface due to wear can be determined by integrating of wear velocity over time from start moment until current time

$$u_{\mathrm{w}}(z,t) = \int_{\tau_0}^{t} \upsilon_{\mathrm{w}}(z,\tau)\,\mathrm{d}\tau = -\frac{k_{\mathrm{w}} V}{H(z)} \int_{\tau_0}^{t} q(z,\tau)\,\mathrm{d}\tau. \tag{1}$$

Based on the results of [7–13], it can be obtained that the displacement of the outer face of the coating, due to the load on the coating outer boundary and pressure on the pipe inner boundary, is represented by the expression

$$u_{\mathrm{q}}(z,t) = \frac{\Theta P_{\mathrm{in}}(t)}{E_{\mathrm{in}}} - \frac{1-\nu_{\mathrm{out}}^2}{E_{\mathrm{out}}} h_{\mathrm{out}}(z,t) q(z,t) - \frac{2(1-\nu_{\mathrm{in}}^2)}{\pi E_{\mathrm{in}}} \int_{-a}^{a} k_{\mathrm{cyl}}\left(\frac{z-\zeta}{r_{\mathrm{in}}}\right) q(\zeta,t)\,\mathrm{d}\zeta,$$
$$h_{\mathrm{out}}(z,t) = h_{\mathrm{out}0}(z) + u_w(z,t). \tag{2}$$

In this expression $q(z,t)$ is contact pressure on outer surface of the coating (or normal load on outer surface on the region $[-a,a]$), E_{in} and E_{out} are Young's moduli of inner pipe and coating, ν_{in} and ν_{out} are

Poisson's ratios of inner pipe and coating, Θ is known coefficient (we use assumptions $h_{out0}(z) \ll \{r_{out}, h_{in}\}$):

$$\Theta = \frac{a_1}{a_2 E_{out}/E_{in} + a_3}, \qquad a_1 = \frac{4(1-\nu_{out})(1-\nu_{in}^2) r_{in}^2 r_{out}}{(r_{out}+\bar{h}_{out})(r_{out}^2 - r_{in}^2)},$$

$$a_2 = \frac{1+\nu_{in}}{1+\nu_{out}} \frac{\bar{h}_{out}[(1-2\nu_{in})r_{out}^2 + r_{in}^2](2r_{out}+\bar{h}_{out})}{r_{out}(r_{out}^2 - r_{in}^2)(r_{out}+\bar{h}_{out})^2}, \quad a_3 = \frac{(1-2\nu_{out})r_{out}}{(r_{out}+\bar{h}_{out})^2} + \frac{1}{r_{out}},$$

$k_{cyl}(s)$ is known kernel of the cylindrical contact problem

$$k_{cyl}(s) = \int_0^\infty \frac{L(u)}{u} \cos(su)\,du,$$

where

$$L(u) = \frac{u[u^2 B_1^2(u) - f(1,u) D_1^2(u) - 1]}{S(u)}, \quad f(r,u) = \frac{2(1-\nu_{in})}{r} + u^2 r, \quad k_r = \frac{r_{out}}{r_{in}},$$

$$S(u) = \frac{f(1,u)}{k_r} + f(k_r,u) + k_r u^4 A_1^2(u) - u^2 f(k_r,u) B_1^2(u) - k_r u^2 f(1,u) C_1^2(u) + f(1,u) f(k_r,u) D_1^2(u),$$

$$A_1(u) = \mathrm{I}_0(u)\mathrm{K}_0(k_r u) - \mathrm{I}_0(k_r u)\mathrm{K}_0(u), \quad B_1(u) = \mathrm{I}_0(u)\mathrm{K}_1(k_r u) + \mathrm{I}_1(k_r u)\mathrm{K}_0(u),$$

$$C_1(u) = \mathrm{I}_0(k_r u)\mathrm{K}_1(u) + \mathrm{I}_1(u)\mathrm{K}_0(k_r u), \quad D_1(u) = \mathrm{I}_1(u)\mathrm{K}_1(k_r u) - \mathrm{I}_1(k_r u)\mathrm{K}_1(u),$$

$\mathrm{I}_0(u)$, $\mathrm{I}_1(u)$, $\mathrm{K}_0(u)$, and $\mathrm{K}_1(u)$ are Bessel functions. Note that $a_2 \ll a_3$ and $E_{out} \lesssim E_{in}$. Hence $\Theta \approx a_1/a_3 \approx 2(1 - \nu_{in}^2) r_{in}^2 r_{out}/(r_{out}^2 - r_{in}^2)$.

Since we assume that wear is much less than the thickness of the coating, then $u_w(z,t) \ll h_{out0}(z)$. Hence expression for displacement due to load (2) can be simplified

$$u_q(z,t) = \frac{\Theta P_{in}(t)}{E_{in}} - \frac{1-\nu_{out}^2}{E_{out}} h_{out0}(z) q(z,t) - \frac{2(1-\nu_{in}^2)}{\pi E_{in}} \int_{=a}^{a} k_{cyl}\left(\frac{z-\zeta}{r_{in}}\right) q(\zeta,t)\,d\zeta. \quad (3)$$

Note that it is linear equation over $q(z,t)$.

Summing displacements due to wear (1) and forces (3), we get following expression for displacement of outer surface of coating

$$u(z,t) = \frac{\Theta P_{in}(t)}{E_{in}} - \frac{1-\nu_{out}^2}{E_{out}} h_{out0}(z) q(z,t) - \frac{k_w V}{H(z)} \int_{\tau_0}^{t} q(z,\tau)\,d\tau - \frac{2(1-\nu_{in}^2)}{\pi E_{in}} \int_{-a}^{a} k_{cyl}\left(\frac{z-\zeta}{r_{in}}\right) q(\zeta,t)\,d\zeta. \quad (4)$$

On the other hand, the displacement of the coating outer surface can be represented as follows

$$u(z,t) = g(z) - [r_{out} + h_{out}(z,t)], \quad (5)$$

where $h_{out}(z,t)$ determined by (2).

Comparing expressions (4) and (5), we obtain the linear mixed integral equation of the problem

$$\frac{1-\nu_{out}^2}{E_{out}} h_{out0}(z) q(z,t) + \frac{2k_w V}{H(z)} \int_{\tau_0}^{t} q(z,\tau)\,d\tau + \frac{2(1-\nu_{in}^2)}{\pi E_{in}} \int_{-a}^{a} k_{cyl}\left(\frac{z-\zeta}{r_{in}}\right) q(\zeta,t)\,d\zeta$$

$$= \frac{\Theta P_{in}(t)}{E_{in}} - g(z) + r_{out} + h_{out0}(z). \quad (6)$$

We will make assumption that due to technological process of coating production its hardness $H(z)$ is greater where coating thickness $h_{out0}(z)$ is less (in areas with less thickness, the surface is hardened better). In this case $H(z) = k_H/h_{out0}(z)$. Unfortunately in the case when these values are independent or the dependence between them is another, it is necessary to construct a numerical solution to the problem (6). So final mixed integral equation has a form

$$\frac{1-\nu_{out}^2}{E_{out}} h_{out0}(z)q(z,t) + \frac{2k_w V h_{out0}(z)}{k_H} \int_{\tau_0}^{t} q(z,\tau)\,d\tau + \frac{2(1-\nu_{in}^2)}{\pi E_{in}} \int_{-a}^{a} k_{cyl}\left(\frac{z-\zeta}{r_{in}}\right) q(\zeta,t)\,d\zeta$$
$$= \frac{\Theta P_{in}(t)}{E_{in}} - g(z) + r_{out} + h_{out0}(z). \quad (7)$$

3. Dimensionless form and solution of the problem

We introduce into (7) new variables and functions by the formulas

$$z^* = \frac{z}{a},\quad \zeta^* = \frac{\zeta}{a},\quad t^* = \frac{t}{\tau_0},\quad P^*(t^*) = \frac{1}{a}\left[\frac{\Theta P_{in}(t)}{E_{in}} + r_{out}\right],\quad g^*(z^*) = \frac{g(z) - h_{out0}(z)}{a},$$

$$m^*(z^*) = \frac{E_{in}}{2(1-\nu_{in}^2)a}\frac{1-\nu_{out}^2}{E_{out}} h_{out0}(z),\quad q^*(z^*,t^*) = \frac{2(1-\nu_{in}^2)q(z,t)}{E_{in}},$$

$$\mathbf{A}^* f(z^*) = \int_{-1}^{1} k_{cyl}^*(z^*,\zeta^*) f(\zeta^*)\,d\zeta^*,\quad \mathbf{V}^* f(t^*) = \int_{1}^{t^*} K^*(t^*,\tau^*) f(\tau^*)\,d\tau^*,$$

$$k_{cyl}^*(z^*,\zeta^*) = \frac{1}{\pi} k_{cyl}\left(\frac{z-\zeta}{r_{in}}\right),\quad K^*(t^*,\tau^*) = -V^* = -\frac{2k_w V_{out}\tau_0}{k_H(1-\nu_{out}^2)}.$$

Then mathematical model for contact-wear problem in dimensionless form

$$m^*(z^*)(\mathbf{I} - \mathbf{V}^*)q^*(z^*,t^*) + \mathbf{A}^* q^*(z^*,t^*) = P^*(t^*) - g^*(z^*). \quad (8)$$

Here **I** is identity operator. It is necessary to determine the function $q^*(z^*,t^*)$ from the obtained equation (8). Note that functions $m^*(z^*)$ and $g^*(z^*)$ in (8) relate with properties and forms of contacting surfaces; these parameters can be described by a rapidly changing function.

Next, we will construct a solution for the dimensionless equation (8), and for brevity and convenience, we will omit the asterisks in notations of constants, variables, functions, and operators.

In integral equation (8), we make new kernel $k(z,\zeta)$ and unknown function $Q(z,t)$

$$Q(z,t) = \sqrt{m(z)}q(z,t) + (\mathbf{I}+\mathbf{R})\frac{g(z)}{\sqrt{m(z)}},\quad k(z,\zeta) = \frac{k_{cyl}(z,\zeta)}{\sqrt{m(z)}\sqrt{m(\zeta)}},$$
$$\mathbf{F}f(z) = \int_{-1}^{1} k(z,\zeta) f(\zeta)\,d\zeta,\quad (\mathbf{I}+\mathbf{R}) = (\mathbf{I}-\mathbf{V})^{-1},\quad \mathbf{R}f(t) = -V\int_{1}^{t} e^{-V(t-\tau)} f(\tau)\,d\tau, \quad (9)$$

where **R** is Volterra operator with kernel $-Ve^{-V(t-\tau)}$ (note that $-Ve^{-V(t-\tau)}$ is kernel of resolvent for Volterra integral equation with kernel $-V$, see [14]). Then equation (8) take a form

$$(\mathbf{I}-\mathbf{V})Q(z,t) + \mathbf{F}Q(z,t) = \frac{P(t)}{\sqrt{m(z)}} + \frac{\tilde{c}(t)\tilde{g}(z)}{\sqrt{m(z)}},\quad z\in[-1,1],\quad t\ge 1, \quad (10)$$

where

$$\mathbf{F}f(z) = \int_{-1}^{1} k(z,\zeta) f(\zeta) d\zeta,\quad \tilde{g}(z) = \int_{-1}^{1} \frac{k_{cyl}(z,\zeta)g(\zeta)}{m(\zeta)} d\zeta,\quad c(t) = (\mathbf{I}+\mathbf{R})1 = e^{-V(t-1)}. \quad (11)$$

We will see the solution of our problem in the class of time-continuous functions in a Hilbert space $L_2[-1,1]$ (e.g., see [15]). To this end we will construct orthonormal basis in $L_2[-1,1]$ from the linearly independent system of functions $\{1/\sqrt{m(z)}, z/\sqrt{m(z)}, z^2/\sqrt{m(z)}, \mathrm{K}\}$ by the formulas

$$J_i = \int_{-1}^{1} \frac{\zeta^i}{m(\zeta)} \mathrm{d}\zeta, \quad d_{-1} = 1, \quad d_i = \begin{vmatrix} J_0 & \Lambda & J_i \\ \mathrm{M} & \mathrm{O} & \mathrm{M} \\ J_i & \Lambda & J_{2i} \end{vmatrix}, \quad p_0(z) = \frac{1}{\sqrt{J_0}}, \quad p_0^{\circ}(z) = \frac{1}{\sqrt{J_0 m(z)}},$$

$$p_j(z) = \frac{1}{\sqrt{d_{j-1} d_j}} \begin{vmatrix} J_0 & J_1 & \Lambda & J_j \\ \mathrm{M} & \mathrm{M} & \mathrm{O} & \mathrm{M} \\ J_{j-1} & J_j & \Lambda & J_{2j-1} \\ 1 & z & \Lambda & z^j \end{vmatrix}, \quad p_j^{\circ}(z) = \frac{p_j(z)}{\sqrt{m(z)}}, \quad i, j = 0, 1, 2, \mathrm{K}, \tag{12}$$

Note that $p_j(z)$ (j = 0, 1, 2, ...) are polynomials. So we will seek the solution in the form

$$Q(z,t) = \sum_{k=0}^{\infty} f_k(t) \Phi_k(z), \tag{13}$$

where $f_k(t)$ is the desired function and the eigenfunctions $\Phi_k(z)$ of the operator $\mathbf{F}$ are determined by solving the spectral problem for the operator $\mathbf{F}$,

$$\mathbf{F}\Phi_k(z) = \gamma_k \Phi_k(z). \tag{14}$$

The eigenfunction system $\{\Phi_k(z)\}_{k=0,1,2,\mathrm{K}}$ is related to the initial system $\{p_k^{\circ}(z)\}_{k=0,1,2,\mathrm{K}}$ of basis functions in $L_2[-1,1]$ as

$$\Phi_k(z) = \sum_{i=0}^{\infty} \psi_i^{(k)} p_i^{\circ}(z). \tag{15}$$

Therefore, solving the spectral problem is reduced to solving the system of algebraic equations

$$\sum_{l=0}^{\infty} R_{il} \psi_l^{(k)} = \gamma_k \psi_i^{(k)}, \quad R_{il} = \int_{-1}^{1}\int_{-1}^{1} \frac{k_{\mathrm{cyl}}(z,\zeta) p_i(z) p_l(\zeta)}{m(z) m(\zeta)} \mathrm{d}z \, \mathrm{d}\zeta, \quad i, l, k = 0, 1, 2, \mathrm{K} \tag{16}$$

We substitute the representation (13) into the integral equation (10) and use (11), (12), (14), and (15) to obtain following expression for the functions $f_k(t)$,

$$f_k(t) = (\mathbf{I} + \mathbf{W}_k) \frac{\psi_0^{(k)} \sqrt{J_0} P(t) + g_k \tilde{c}(t)}{1 + \gamma_k}, \quad g_k = \sum_{i=0}^{\infty} \psi_i^{(k)} \sum_{l=0}^{\infty} R_{il} \int_{-1}^{1} \frac{p_l(\zeta) g(\zeta)}{m(\zeta)} \mathrm{d}\zeta,$$

where $\mathbf{W}_k$ are the Volterra operators whose kernels $R_k(t,\tau) = -\frac{V}{1+\gamma_k} \exp[-\frac{V(t-\tau)}{1+\gamma_k}]$ are resolvents of the kernels $-V/(1 + \gamma_k)$. Then the functions $f_k(t)$ can be represented as

$$f_k(t) = \frac{\psi_0^{(k)} \sqrt{J_0}}{1 + \gamma_k} \left\{ P(t) - \frac{V}{1 + \gamma_k} \int_1^t \exp\left[-\frac{V(t-\tau)}{1+\gamma_k} \right] P(\tau) \mathrm{d}\tau \right\} + \frac{g_k}{\gamma_k} \left\{ e^{-V(t-1)} - \frac{1}{1+\gamma_k} \exp\left[-\frac{V(t-1)}{1+\gamma_k} \right] \right\}. \tag{17}$$

As a result, by using (9), (11)–(13), (15), we finally obtain the expression for the contact stresses

$$q(z,t) = \frac{1}{m(z)} \sum_{k=0}^{\infty} f_k(t) \sum_{m=0}^{\infty} \psi_m^{(k)} p_m(z) - e^{-V(t-1)} \frac{g(z)}{m(z)},$$

where the functions $f_k(t)$ are determined by relation (17), the coefficients $\psi_l^{(k)}$ can be obtained from (16), and $p_k(z)$ can be calculated by the formulas (12).

4. Conclusions and remarks

Contact-wear problem for elastic tube with rough coating and rigid bush with complex inner shape is posed and solved. Analytical representation for contact stresses as a series over special basis is obtained. Complex functions connected with surface shapes and coating hardness is represented by separate factors and terms. It allow one to provide effective numerical calculations using a small number of members of the series. It is important in the case when characteristics of coating and bush described by rapidly changing functions. Another known approaches (using Legendre polynomials, trigonometric functions, etc.) lead us to computational errors.

Obtained representation of contact pressures provide solution only for the case when wear is much less then coating thickness. Another important assumptions described in the article. But this solution can be first step for modeling much more complicated cases of wear.

Acknowledgments

The present work was partially supported by the Ministry of Science and Higher Education within the framework of the Russian State Assignment (contract No. AAAA-A20-120011690132-4) and partially supported by Russian Foundation for Basic Research (projects Nos. 18-01-00770 and 19-51-60001).

References

[1] Alexandrov V M and Mkitaryan S M 1979 *Contact Problems for Bodies with Thin Coatings and Interlayers* (Moscow: Nauka) [in Russian]
[2] Pronikov A S 1957 *Wear and Durability of Machines* (Moscow: Mashgiz) [in Russian]
[3] Khrushchev M M and Babichev M A 1970 *Abrasive Wear* (Moscow: Nauka) [in Russian]
[4] Collins J 1993 *Failure of Materials in Mechanical Design: Analysis, Prediction, Prevention* (New York: Wiley)
[5] Goryacheva I G and Dobychin M N 1988 *Contact Problems in Tribology* (Moscow: Mashinostroenie) [in Russian]
[6] Soldatenkov I A 2010 *Wear Contact Problem with Applications to Engineering Calculation of Wear* (Moscow: Fizmatkniga) [in Russian]
[7] Manzhirov A V and Chernysh V A 1988 On the interaction of a rigid reinforcing sleeve and inhomogeneous aging high-pressure pipes *Mech. Solids* **22** (6) 104–10
[8] Arutyunyan N Kh and Manzhirov A V 1999 *Contact Problems in the Theory of Creep* (Yerevan: Izd-vo Inst. Mekhaniki NAN Armenii) [in Russian]
[9] Manzhirov A V and Kazakov K E 2017 Axisymmetric contact between a rigid punch and a coated foundation with rough surfaces *J. Phys.: Conf. Ser.* **937** 012028 doi: 10.1088/1742-6596/937/1/012028
[10] Manzhirov A V and Kazakov K E 2018 Axisymmetric problem of fretting wear for a foundation with a nonuniform coating and rough punch *AIP Conf. Proc.* **1959** 070023 doi: 10.1063/1.5034698
[11] Manzhirov A V and Kazakov K E 2018 Modeling the contact interaction between a nonuniform foundation and a rough punch *Math. Models Comput. Simul.* **10** (3) 314–21 doi: 10.1134/S2070048218030109
[12] Kurdina S P and Kazakov K E 2019 Axisymmetric contact problem for a punch and nonuniform foundation with rough surfaces *AIP Conf. Proc.* **2116** 380006 doi: 10.1063/1.5114387
[13] Kazakov K E 2019 Wear of foundation with a nonuniform coating by rough punch *IOP Conf. Ser.: Mater Sci. Engng* **489** 012027 doi: 10.1088/1757-899X/489/1/012027
[14] Polyanin A D and Manzhirov A V 2008 *Handbook of Integral Equations*, 2nd edition (Boca Raton: Chapman & Hall/ CRC)
[15] Manzhirov A V 2016 A mixed integral equation of mechanics and a generalized projection method of its solution *Dokl. Phys.* **61** (10) 489–93 doi: 10.1134/S1028335816100025

IOP Conf. Series: Materials Science and Engineering **904** (2020) 012011 doi:10.1088/1757-899X/904/1/012011

Computational-experimental model of fatigue growth of the welded joint crack

Mikhail S. Kuts, Ludmila A. Andrienko, Phillip M. Annenkov, Yaroslav V. Alexandrov

Bauman Moscow Technical University, 105005, ul. Baumanskaya 2-ya, 5, Moscow, Russia

E-mail: kuts@bmstu.ru

Abstract. Welded joints are commonly used in various industries, and in many cases, they contain an initial crack, which could lead to huge damage. For this reason, it is essential to have a model to predict fatigue evolution of the welded joint crack. This paper is dedicated to the method of constructing the computational-experimental model of the fatigue growth of the welded joint crack. The described method is based on a numerical calculation using finite element modelling and Paris-Erdogan equation of the crack growth. Besides, the fatigue experiments of the test sample with a welded joint were conducted and the model of crack growth was constructed with the use of the obtained experimental data.

1. Introduction

Welded joints are widely used in all industries from automotive to aerospace industry. At the same time, a lot of welded joints contain local defects, which is hardly identified and could lead to construction damage [1, 2]. Apart from that, it is important to diagnose a growth of the crack and to predict its critical condition for timely service.

One of the most common diagnostic methods is an experimental modal analysis by means of which we could find changes in construction state by changes in its dynamic response [3, 4]. To achieve it we need to know a relation between the changes in structure dynamic response and its residual life [5], which could be found by fatigue tests. A problem of the fatigue test is its duration. An acceleration of the tests can be reached by increasing a load frequency, which could be made by resonance testing on a shaker. The disadvantage of such testing is unstable loading caused by changing of construction stiffness and damping, which leads to computational difficulties.

In this paper, the method of conducting such tests using finite element (FE) simulation of resonance vibrations of tested samples and obtained computational-experimental model of fatigue growth of the welded joint crack are described.

2. Description of experimental apparatus and methods used

2.1. Investigated object

In this work we investigated the growth of initial crack in welded joint in steel beam. Beams used in experiment are shown in Figure 1. Beams were made from two steel strip, which the width and thickness of 30 mm and 4 mm, accordingly, with the use of single butt-joint seam. Further, beams were clamped by the short strip on the experimental stand and the long strip hangs as cantilever so the butt joint was on its base.

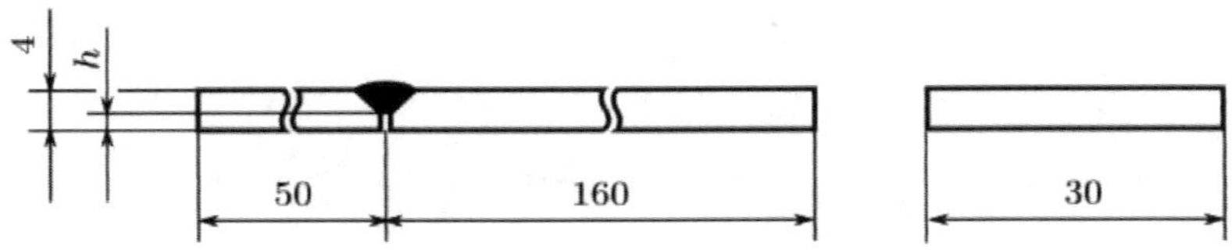

Figure 1. Beam used in experiments.

On the other side of the seam there was the lack of fusion, which had various length l_{init}. The lack of fusion was an initial crack, which was growing during the experiment.

2.2. Experimental stand

Experiments were held on the SignalForce electrodynamic shaker by Data Physics. The tested beams were fixed on the shaker by a rigid holder with accurate machined contact surfaces to reduce influence of the bolted joint on dynamic behaviour of the system [6]. Tests were conducted in the resonance for increasing of the beam response amplitude. Test samples were designed with the first natural frequency near 120Hz to reduce a testing time.

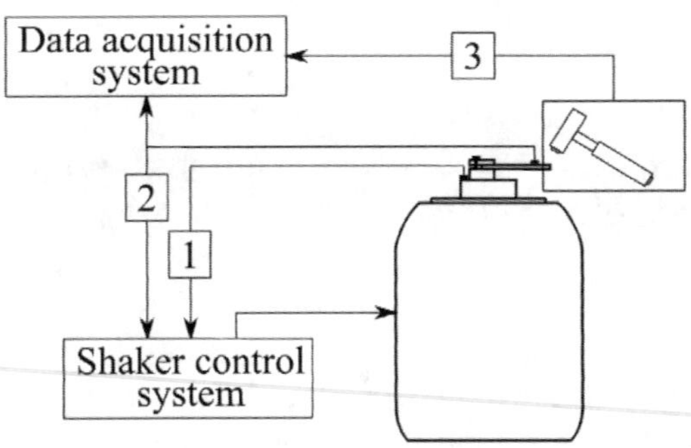

Figure 2. Scheme of the experimental stand: 1 – shaker excitation feedback; 2 – beam response feedback; 3 – hammer excitation feedback.

Figure 3. Shaker with mounted test sample.

To acquire the data, such as beam acceleration and excitation frequency, NI USB-4431 module was used with program written in LabView. The flow chart of the acquisition program algorithm is shown in Figure 4. Also an impact hammer to measure the force excitation feedback for additional modal testing was also provided there.

2.3. Stress intensity factor and crack growth rate

Accordingly to the Paris-Erdogan law [7, 8, 9], the evolution of crack could be obtained as the solution of differential equation:

$$\frac{dl}{dN} = C(\Delta K)^m, \tag{1}$$

IOP Conf. Series: Materials Science and Engineering **904** (2020) 012011 doi:10.1088/1757-899X/904/1/012011

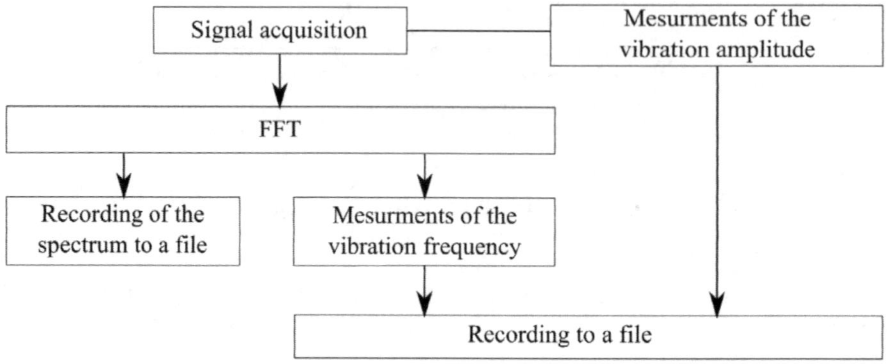

Figure 4. Flow chart of the acquisition program algorithm.

where l — a crack length, $\Delta K = K_{max} - K_{min}$ — an amplitude of the stress intensity factor, which depends on a stress in the crack area and the crack length, C and m — empirical coefficients for the certain material.

The investigated beam could be considered as flat plate under the bending moment with crack of a 1^{st}mode [7]. In case of loading by the periodically changing bending moment, the minimum of the stress intensity factor will be equal to zero and its amplitude could be calculated by the following equation [10]:

$$K_I = F_I(\alpha)\frac{6M}{wh^2}\sqrt{\pi l} \tag{2}$$

where M is an amplitude of the bending moment in the crack area, w and h are strip width and thickness correspondingly, $F_I(\alpha) = 1.122 - 1.40\alpha + 7.33(\alpha)^2 - 13.08(\alpha)^3 + 14.0(\alpha)^4$ for $\alpha = \frac{l}{h}$.

2.4. Vibration simulation using FEM

To calculate the bending moment we used a finite elements method in the statement of flat dynamical problem with beam elements. According to the [11, 12] vibration of the body under harmonic load with circular frequency ω can be described by differential equation

$$\mathbf{M}\ddot{\mathbf{u}} + \mathbf{C}\dot{\mathbf{u}} + \mathbf{K}\mathbf{u} = \mathbf{f}e^{i\omega t}, \tag{3}$$

where $\mathbf{M}, \mathbf{C}, \mathbf{K}$ – global mass, damping and stiffness matrices of the body, $\mathbf{u}$ – vector of displacements. Assuming that the solution of the (3) is limited by the excitation force frequency, we will find it in a form

$$\mathbf{u}(t) = \sum_j v_j \mathbf{v}_j e^{i\omega t}, \tag{4}$$

where $\mathbf{v}_j$ is a normalized vector of displacements corresponded to j^{th}body eigenform (or mode shape) and v_j — its corresponding amplitude. Eigenvectors could be approximately obtained from eigendecomposition of the matrix $\mathbf{A}$ [13], which is

$$\mathbf{A} = \mathbf{M}^{-1}\mathbf{K}. \tag{5}$$

The amplitude of jth eigenform could be found from solution of (3) or in case of uncertain damping matrix $\mathbf{C}$ amplitudes v_j it would be easier to find it directly from known response data:

$$v_j = \frac{a_{z_0}}{\omega^2 \mathrm{v}_j(z_0)}, \tag{6}$$

IOP Conf. Series: Materials Science and Engineering **904** (2020) 012011 doi:10.1088/1757-899X/904/1/012011

where a_{z_0} —acceleration of the measured point with coordinate z_0, $\mathrm{v}_j(z_0)$ — displacement of that point corresponded to the jth normalized eigenform in that point.

Vector of displacements of the certain point of the body is calculated by:

$$\mathrm{v}_j(z) = \begin{pmatrix} 1 & 0 \end{pmatrix} \mathbf{N}^e \mathbf{v}_j^e \tag{7}$$

where $\mathbf{N}^e$ — form function matrix of the eth element and $\mathbf{v}_j^e$ is a local displacements vector of jth normalized eigenform, components of which are corresponding to certain components of $\mathbf{v}_j$. Computation of local displacements vector $\mathbf{u}^e$ from the global displacements vector $\mathbf{u}$ could be written as

$$\mathbf{u}^e = \mathbf{T}\mathbf{u}, \tag{8}$$

where $\mathbf{T}_{n\times m}$ is translate matrix with m - number of allowed element displacements and m - number of allowed body displacements. In (8) vector $\mathbf{u}$ could be changed by $\mathbf{v}_j$.

The element stiffness and mass matrices can be evaluated by following equations

$$\mathbf{k}^e = \frac{EI}{l^3}\begin{pmatrix} 12 & 6l & -12 & 6l \\ 6l & 4l^2 & -6l & 2l^2 \\ -12 & -6l & 2 & -6l \\ 6l & 2l^2 & -6l & 4l^2 \end{pmatrix} \quad \text{and} \quad \mathbf{m}^e = \frac{\rho Al}{420}\begin{pmatrix} 156 & 22l & 54 & -13l \\ 22l & 4l^2 & 13l & -3l^2 \\ 54 & 13l & 156 & -22l \\ -13l & -3l^2 & -22l & 4l^2 \end{pmatrix},$$

where E — Young's modulus of the beam material, ρ — density of the beam material, A — cross section area, I — cross section moment of inertia, l — element length. Then the global stiffness matrix can be simply evaluated as

$$\mathbf{K}_{n\times n} = \sum_{e=1}^{N} \mathbf{K}^e_{n\times n}, \tag{9}$$

where $\mathbf{K}^e_{n\times n}$ — local stiffness matrix $\mathbf{k}^e$ extended to the global one dimension, N — number of elements in mesh. The global mass matrix van be evaluated (assembled) by the same way.

Next step is calculation of the internal bend moment, caused by dynamical bending of the beam. The internal moment of bent beam is

$$M(z) = EI\frac{\partial^2 u}{\partial z^2} \tag{10}$$

The second derivative of the displacements for each element could be found as

$$\left.\frac{\partial^2 u}{\partial z^2}\right|_{z\in L^e} = \mathbf{B}\mathbf{u}^e \tag{11}$$

where $\mathbf{B}$ — matrix of deformations of the element and $\mathbf{u}^e$ is an elements displacements vector, which components are corresponding components of the global displacements vector $\mathbf{u}$. Accordingly, the bending moment will be simply computed by substitution of (11) into (10).

3. Tests results

3.1. Obtained experimental data

An example of obtained response amplitudes and frequencies time data is presented in Figure 5. Since the system damping had been increasing during the experiments due to crack growth, the response amplitude was decreased during the tests and we were needed to increase the amplitude of excitation to hold the high crack growth rate. For this reason, the test was hold by on five parts, each of them has a different excitation signal level (Figure5–*a*). And as a result, the total amplitude-time relation has a chaotically changing character.

Furthermore, as it is shown in Figure 5–*b*, the frequency changes in the stair-step manner due to the low accuracy of frequency identification methods used. Its makes us impossible to calculate the crack growth rate directly from the experiments.

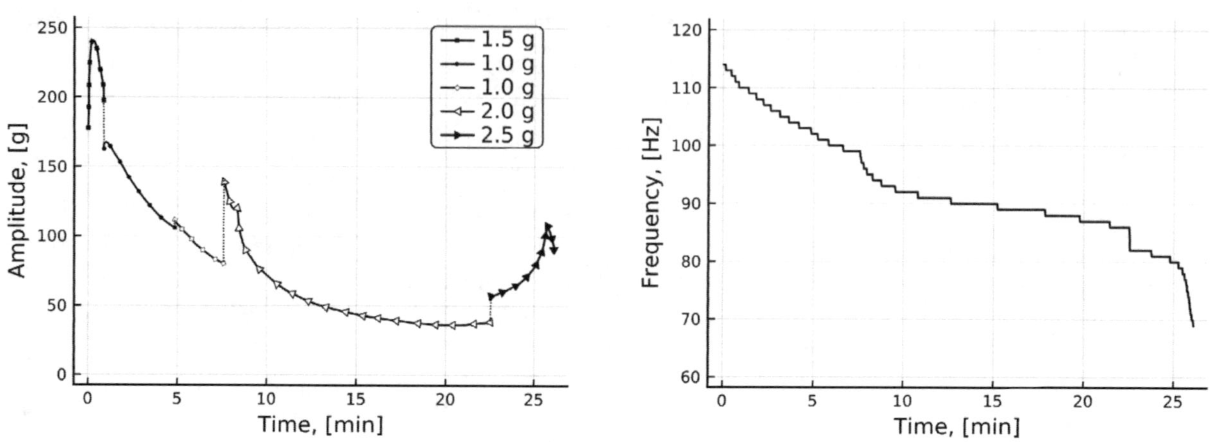

a) *b)*

Figure 5. Example of amplitude (*a*) and resonance frequency evolution (*b*) during the test.

Table 1. Identified model parameters.

Parameter	Beam no. 1	Beam no. 2
C	1.024^{-11}	1.063^{-11}
m	3.35	3.3
l_{init}, mm	1.4	1.2

3.2. Preparing of the data and computation algorithm

First step in the calculation is replacement of time variable t in obtained experimental sequences $a_{max}(t)$ and $f_r(t)$ (which equal to the first natural frequency) to cycle numbers N. This replacement could be made by the following:

$$N(t) = \int_0^t f_r(\tau)d\tau. \tag{12}$$

Further we can solve the DE (1) using Euler methods, so the crack length evolution would be found in iterative calculation:

$$\begin{cases} l_i = l_{i-1} + dl_{i-1} = l_{i-1} + C\left(K_I(a_{i-1}, li-1)^m\right)dN_{i-1} \\ l_0 = l_{init} \end{cases} \tag{13}$$

There are three unknown parameters in (13) — C, m and in general l_{init} — which could be identified.

After the crack length calculated we can calculate natural frequency of the beam. The relation between crack length l and the first natural frequency of the beam f_1 was obtained by parametric numerical modal analysis of flat beam with transverse crack, with use of ANSYS Workbench. The mesh used is shown in Figure 6. The elements PLANE183 are used for meshing. Size of the elements was approximately 0.2 mm in the crack area and 1 mm beyond the crack. The overall number of the elements was 3092. The obtained relation is plotted in Figure 7.

IOP Conf. Series: Materials Science and Engineering **904** (2020) 012011 doi:10.1088/1757-899X/904/1/012011

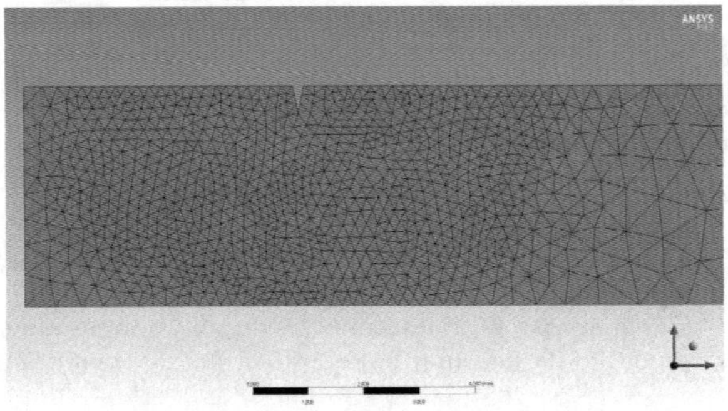

Figure 6. Finite element mesh.

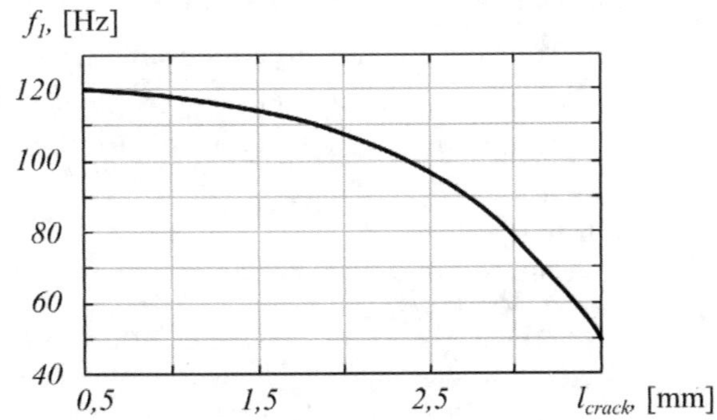

Figure 7. Relation between crack length l and first natural frequency of the beam f_1.

3.3. Results

The calculated evolution of the crack length and caused changes in natural frequency during the destruction process are shown in Figure 8. Values of identified parameters i Paris-Erdogan model (1) presented in table 1. In the current work these parameters were identified manually.

As we can see in Figure 8-*b*, proposed method gives a good propagation of the first natural frequency evolution, except the last curve section, where curve character remains, but quantitative values give an error. According to the authors opinion, this effect can be explained by an increased non-linearity of vibration of the strip with large crack.

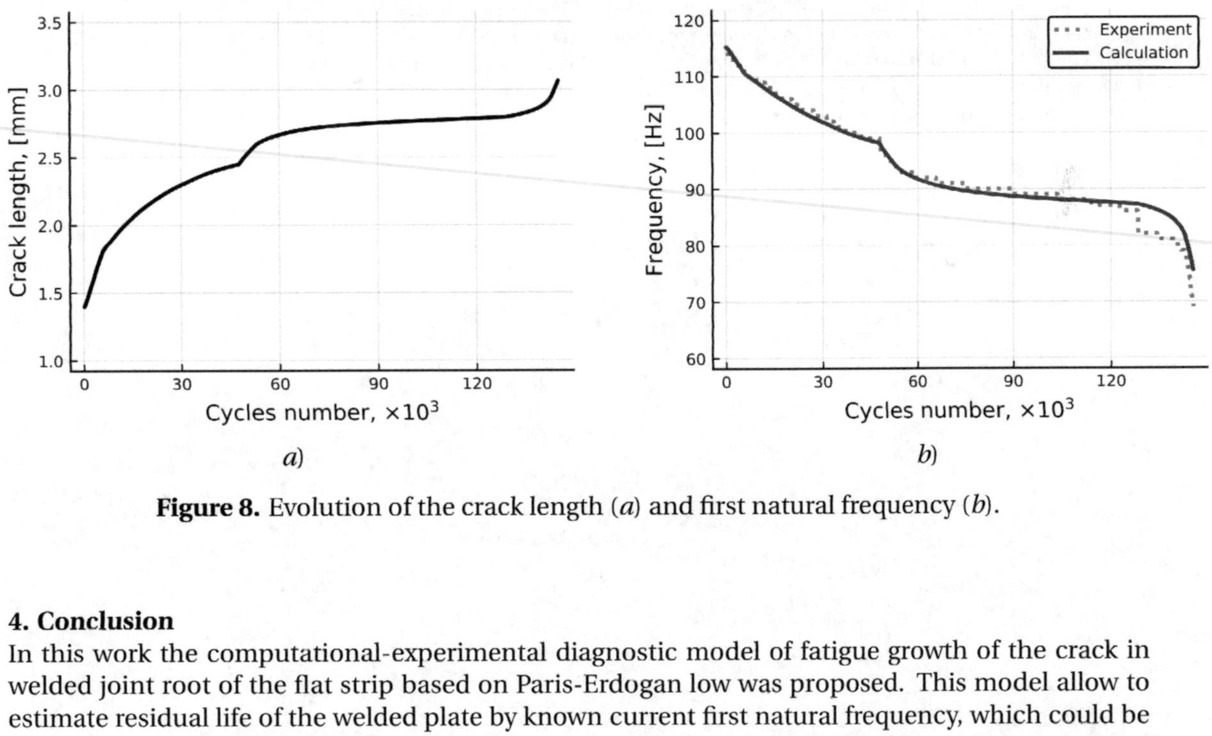

Figure 8. Evolution of the crack length (*a*) and first natural frequency (*b*).

4. Conclusion

In this work the computational-experimental diagnostic model of fatigue growth of the crack in welded joint root of the flat strip based on Paris-Erdogan low was proposed. This model allow to estimate residual life of the welded plate by known current first natural frequency, which could be obtained by the experimental modal analysis of the part. Calculation of crack growth shows a good convergence with the experimental data.

It is planned to proceed further works in three directions:

IOP Conf. Series: Materials Science and Engineering **904** (2020) 012011 doi:10.1088/1757-899X/904/1/012011

(i) Develop the method to determine the model parameters by means of optimization;
(ii) Consider the non-linear behaviour of the body with large crack under the cyclic loading;
(iii) Consider a non-linear fatigue crack growth, which could be described by the more complicated models instead of Paris-Erdogan model.

References

[1] Zakharov M N, Morozov E M and Nasonov V A 2015 Assessing the Risk of Welding Defects *Russian Engineering Research* **35/11** pp 846–849

[2] Makhutov N A, Pokrovskii A M and Dubovitskii E I 2019 Analysis of Crack Resistance of an Oil Trunk Pipeline Considering the Varying Failure Viscosity in the Neighborhood of a Welded Joint *Journal of Machinery Manufacture and Reliability* **48/1** pp 35–42

[3] Geradin M and Rixen D J 2015 *Mechanical vibrations: theory and application to structural dynamics* (John Wiley & Sons) p 598

[4] Zhulev V and Kuts M 2018 Contact models verification by the finite element model updating method based on the calculation of the sensitivity coefficient *MATEC Web of Conferences* pp 02008

[5] Akulenko L D, Gavrikov A A, and Nesterov S V 2018 Natural Vibrations of a Liquid-Transporting Pipeline on an Elastic Base *Mechanics of Solids* **53/1** pp 101–110

[6] Kuts M and Zhulev V 2019 Two approaches for determination of contact layer stiffness *Proceedings of the 8th IOMAC - International Operational Modal Analysis Conference* Kopenhagen

[7] Pestrikov V M and Morozov E M 2012 *Mekhanika razrushenija. Konspect lektsij (Fracture mechanics. Lectures)* (St.-Petersburg: Professija) p 552

[8] Chernyatin A S, Matvienko Y G and Razumovsky I A 2018 Fatigue Surface Crack Propagation and Intersecting Cracks in Connection With Welding Residual Stresses *Fatigue & Fracture of Engineering Materials & Structures* **41/10** pp 2140–2152

[9] Pokrovskii A M, Dubin D A and Vdovin D S 2019 Fatigue-Crack Propagation in Torsional Shafts Within the Suspension of High-Speed Caterpillar Vehicles *Russian Engineering Research* **39/7** pp 548–555

[10] Murakami Y, Aoki S, Hasebe N, Itoh Y, Miyata H, Miyazaki N, Terada H, Tohgo K, Toya M and Yuuki R 1987 *Stress intensity factors handbook* vol 1 (Pergamon Press) p 1464

[11] Felippa C A 2004 *Introduction to Finite Element Methods* (University of Colorado) p 620

[12] Kim N H 2014 *Introduction to Nonlinear Finite Element Analysis* (N.Y.:Springer) p 430

[13] Zienkiewicz O C , Taylor R L and Zhu J Z 2005 *The Finite Element Method: Its Basis and Fundamentals* (Oxford: Elsevier Butterworth-Heinemann) p 752

IOP Conf. Series: Materials Science and Engineering **904** (2020) 012012 doi:10.1088/1757-899X/904/1/012012

Time-series classification for industrial applications: a brake pad wear prediction use case

Evgeny Burnaev
Skoltech, Moscow, Russia
E-mail: e.burnaev@skoltech.ru

Abstract.
Brake system is an important part for control of a vehicle. Hence condition monitoring of brake pads is essential for ensuring passenger's safety. Many existing methods for brake pads wear assessment rely on specific sensors installed in the brake system, which could be expensive. Instead we use data from existing vehicle's sensors and electronic control unit that are readily available in modern vehicles. We reduced the prediction problem to time-series classification problem and developed and tested several classification pipelines based on machine learning. We demonstrated that it is possible to predict a brake pad wear with an accuracy sufficient for real-life usage.

1. Introduction

1.1. Brake pad wear prediction problem

Brake pads is an important part of a vehicle brake system. Hence condition monitoring of the brake pad is essential for passenger's safety [1]. The brake pad wear can be predicted through identification of some symptoms, manifestated in measurements from specific sensors [2, 3], using approaches for predictive analytics [4] based on failure prediction [5, 6], anomaly detection [7, 8, 9, 10, 11] and time-series change-point detection [12, 13, 14, 15, 16, 17, 18].

However, mounting of the sensors could be expensive and not possible for vehicles used everyday. Thus the only way to predict brake pad wear is to use data from existing vehicle's sensors and electronic control unit that are readily available in modern vehicles [1]. In this setup the problem can be reduced to time-series classificatoin problem: during a brake event we observe signals from sensors and should predict the condition of the brake pad.

There exist a number of time-series classification approaches [19, 20]. However, many of these modern approaches exploit deep neural networks, especially convolutional neural networks, for end-to-end time series classification. Although being automatic these approaches lead to construction of non-interpretable black-boxes that often are prohibited in engineering applications, especially when the final aim is deployment of the predictive model in a real production system. Moreover, the resulting predictive models often are too heavy to be used in on-board systems.

To develop a robust and interpretable predictive system in this work we proposed a time-series classification pipeline based on data preprocessing and feature engineering combined with construction of time-series classifiers capable to utilize historic information about previous braking events. We demonstrated that it is possible to predict a brake pad wear with an accuracy sufficient for real-life usage.

The paper has the following structure. In Section 2 we discuss the proposed approach. In Section 3 we present obtained results, and in Section 4 we draw conclusions.

1.2. Mathematical problem statement

We consider a time-series classification problem statement. We denote by $\mathcal{D} = (\mathbf{X}, \mathbf{y}) = \{(\mathbf{x}_1, y_1), \dots, (\mathbf{x}_N, y_N)\}$ a dataset, where $\mathbf{x}_k \in R^{d \times n_k}$ is a d-dimensional time series of length n_k, and $y_k \in \{1, \dots, K\}$ is a K-class label of the time series. The task is to construct the model that for a new unseen time series $\mathbf{x}_*$ predicts its label.

In practice, the time series can be provided in a chronological order, that is $\mathbf{x}_k$ is observed earlier than $\mathbf{x}_m$ if $k < m$ for $\forall k, m = \overline{1, N}$, and the classes $\{1, \dots, K\}$ are ordered as well. E.g., in case of a brake pad wear prediction in the beginning brake pads are new, then they become used and later they wear out, i.e. labels (`New`, `Used` and `WornOut`) naturally appear in a chronological order [1]. It means that to predict label for a current time series we can use several previous time series (history) to make a predictive model more accurate. Moreover, we also can utilize that the label `Used` can appear only after `New` and the label `Worn Out` appears only after `Used`.

1.3. Real-world data set description

Data set consists of $\sim$ 1500 braking events with brake pad usage condition label (`New`, `Used`, `WornOut`), see [1]. Each braking event is described by a multivariate time series of different signals measured during braking. Some of them have straightforward physical meaning (e.g. acceleration, friction coefficient, pressure and so on), i.e. signals from sensors, and some of them are software parameters recorded by the electronic control unit (ECU) of the braking system. Overall number of parameters is equalt to $\sim$ 400, with length of each time-series ranging from $\sim$ 100 to $\sim 10^3$. There are highly correlated parameters, e.g. characteristics that were measured for each wheel separately. An example of time series is given in Fig. 1 for different brake pads condition.

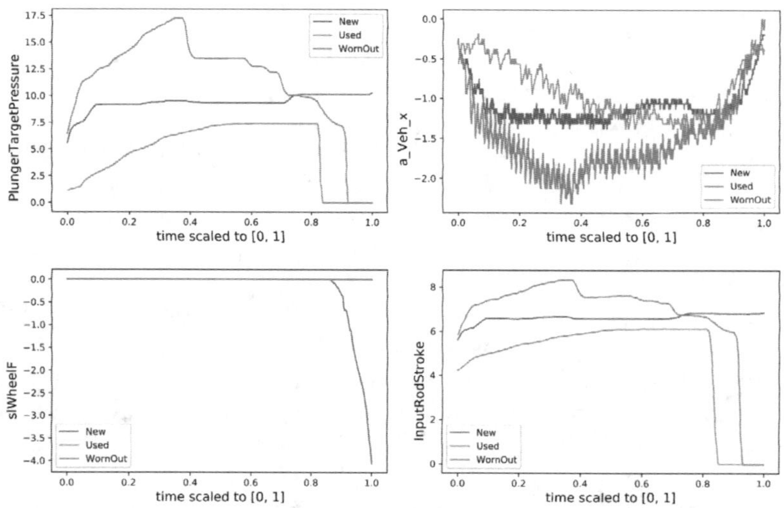

Figure 1. Example of time series

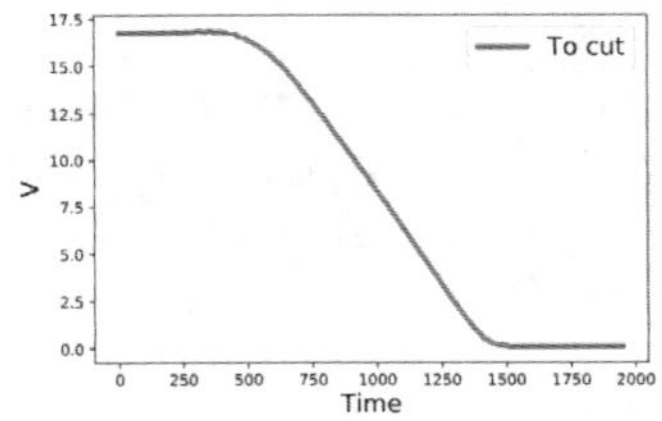

Figure 2. Example of cut time series

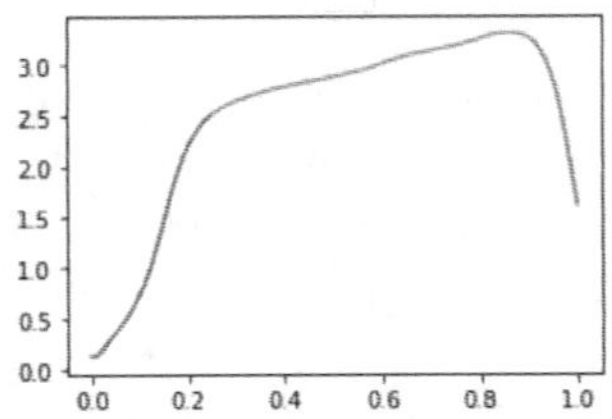

Figure 3. Example of adaptive split.

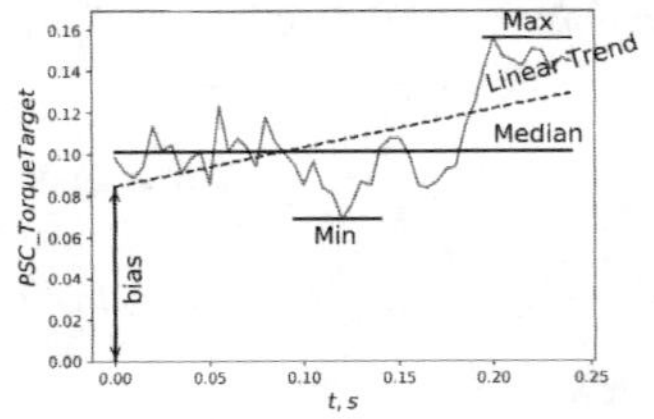

Figure 4. Illustration of local features.

2. Proposed Approach

Model construction algorithm consists of the following steps:

1. Preprocessing (cleaning data, removing unimportant columns, etc.).
2. Feature extraction (calculate new features, extract some statistics from time series, etc.).
3. Build a predictive model.

2.1. Preprocessing

Preprocessing of the dataset consists of the following steps

(i) Remove almost constant parameters: some parameters don't change their values for 99% of braking events.
(ii) Average measurements that correspond to the same parameter but for different wheels.
(iii) Drop some columns that were marked as unimportant by classifier (see below).
(iv) Cut beginning and end of the time series.
(v) Drop observations with constant speed.

Now let us describe each step of preprocessing in more detail.

Removing constant parameters. To identify constant parameters we do the following

(i) For a given parameter calculate its standard deviation for each braking event.
(ii) Calculate number of events for which the standard deviation is close to 0.
(iii) If number of such events is greater than 99% of events mark the parameter as constant.

Averaging parameters over wheels. Some parameters were measured separately for each wheel. For such parameters we do the following

(i) Calculate average over different wheels.
(ii) Put the average to the data set and remove separate measurements.

Cut beginning of time series. A criterion to cut is $\mathrm{RollingStd}_{50}(V) \le \frac{1}{5}\max_V\{\mathrm{RollingStd}_{50}(V)\}$, where V is the velocity and $\mathrm{RollingStd}_{50}$ is a rolling standard deviation of V with window size 50 (in time ticks, where time tick is the interval between two consequent measurements). This criterion is used to cut all time series of a braking event (i.e. the same split points are chosen for all time series of a braking event). All the constants here are chosen manually, just by evaluating the performance quality of the result: we need to cut the beginning of the time series where the braking doesn't begin.

Cut end of time series. A criterion to cut is $\text{RollingStd}_{50}(V) \leq \frac{1}{10}\max_V\{\text{RollingStd}_{50}(V)\}$ and $V < 0.2$. This criterion is used to cut all time series of a braking event (i.e. the same split points are chosen for all time series of a braking event). All the constants here are chosen manually, just by evaluating the performance quality of the result: we need to cut the end of the time series where the braking has already ended. Example of cut time series is given in Figure 2.

Drop observations with constant speed. It turned out that after preprocessing there are braking events with constant speed. We find such events by checking standard deviation of the velocity. If it is close to zero, then we drop the observation from the data set. After this procedure 1290 observations left.

2.2. Feature extraction
Standard classification models require feature vectors of fixed dimension as input. Therefore we need to extract feature vectors from time series characterizing their properties relevant for the classification problem.

2.2.1. Global and local features. We distinguish two types of features: global and local.

Global features. By global features we mean features that are calculated using the whole time series, so they contain some aggregated information about the whole signal. We calculate the following global features

- Absolute energy. Let $\mathbf{x}_i, i = 1, \ldots, T$ be the signal. Then the absolute energy is defined as $\text{abs\_energy} = \sum_{i=1}^{T} \mathbf{x}_i^2$. Actually absolute energy is an area under the square of absolute value of the signal over time.
- Binned entropy. First we bin the signal values into K bins. Then we compute entropy $\text{binned\_entropy} = -\sum_{k=1}^{K} p_k \log p_k \mathbf{1}_{p_k>0}$, where p_k is percentage of values in the k-th bin. It is a measure of disorder or uncertainty in signal. Higher values indicates higher uncertainty (or complexity) of the signal. We set $K = 10$.

Local features are statistics that we calculate on some intervals of the time series.

To identify reasonable intervals we make adaptive split for each braking event as follows. Typically, absolute values of acceleration increases rapidly at the beginning of the braking, after that it remains more or less constant for some time and in the end it rapidly decreases (see example of absolute value of acceleration in Figure 3). That is why we split the time series into three intervals. Split points remain the same for all time series inside the braking event.

For each brake we find two split points such that each of three intervals are well approximated by a line. As a result the braking event is represented by 3 multi-dimensional but shorter time series. In more details the procedure is performed using brute force, i.e.

(i) Choose 2 split points on time series, construct approximation in each of the 3 intervals and calculate total mean squared error.

(ii) Do this for each pair of points and choose the best one in terms of mean squared error.

Additionally, we impose some constraints on the length of intervals, namely, we restrict the intervals to be longer than 5% of total length of time series. The adaptive split can be speeded up by using dynamic programming (details can be found in [21]). The result of adaptive split is 3 intervals in each of which the time series is more or less homogeneous. Extracting features from homogeneous regions allows to obtain more accurate classification.

In each interval following statistics are calculated (see Figure 4): median, standard deviation, maximum, minimum, linear trend: slope, bias, error of approximation. As a result we obtain

IOP Conf. Series: Materials Science and Engineering **904** (2020) 012012 doi:10.1088/1757-899X/904/1/012012

a dataset of size 575 features: after preprocessing 25 parameters left, for each parameter we calculate 2 global features and 21 local features (7 features for each of 3 intervals).

2.3. Models

In this section 2 models are described. The first model corresponds to the first problem statement, where we don't take into account chronological order of the events. We will refer to this type of models as *models without history* or the *first level models*.

The second type of models correspond to the second problem statement where we take into account time stamps of the measurements. These models are based on the first level models: it takes predictions of the first level model (or some features extracted from the model) and constructs another model on top of them (the *second level model*). We will also refer to this type of models as *models with history*.

2.3.1. Models without history. Model construction is performed in three steps

(i) Fit Gradient Boosted Decision Trees model (using XGBoost package).

(ii) Calculate feature importances and select several the most important features.

(iii) Refit model using selected features.

Here decision tree is a model that tries to make prediction by learning simple decision rules inferred from the data; Gradient Boosted Decision Tree is an ensemble of decision trees, so that decision trees are constructed iteratively one-by-one and each new decision tree tries to correct approximation error of already constructed ensemble. Details on Gradient Boosted Decision Tree and its implementation in XGBoost package can be found in [22].

Feature Importance can be measured using *Gain*. Each split of variable improves loss on the training set. The sum of improvements over all splits of variable and all trees in the ensemble is called Gain. The higher the Gain the more important is the variable. In this study we supposed that variable is important if $Gain > 10$ for this variable.

XGBoost model has a lot of hyper-parameters that allows to fine-tune the model. We used the following set of parameters: we set maximum depth of each decision tree in the ensemble `max_depth` to 3, as we keep this value small in order to obtain more robust model; we set a number of trees in the ensemble `n_estimators` to 400, as according to our experiment larger number of trees gives only a marginal improvement (if gives it all) but makes the model much more complex; we set a weight of each newly constructed tree in the ensemble `learning_rate` to 0.05 as reducing learning rate allows to avoid overfitting at the cost of slower training and larger number of trees.

2.3.2. Models with history. Brake pad is not worn out immediately. It is a continuous process, therefore, if during several brakings brake pads are not worn out then with high probability they are not worn out at current braking. This means that we can use several consecutive brakings to improve prediction for current one. It is supposed that we already have the first level model (model without history). Then we can do the following procedure:

1. For braking events $t, t-1, \ldots t-k+1$ calculate probabilities $\mathbf{p}_i = (p^1_{t-i}, p^2_{t-i}, p^3_{t-i})$ $\forall i = \overline{0, k-1}$ using the first level model, where p^c_j is a probability of class c for braking event j.
2. Use the probabilities $\mathbf{p}_i$ as features for the second level model to make predictions (see Figure 5).

For the second level we considered the following models.

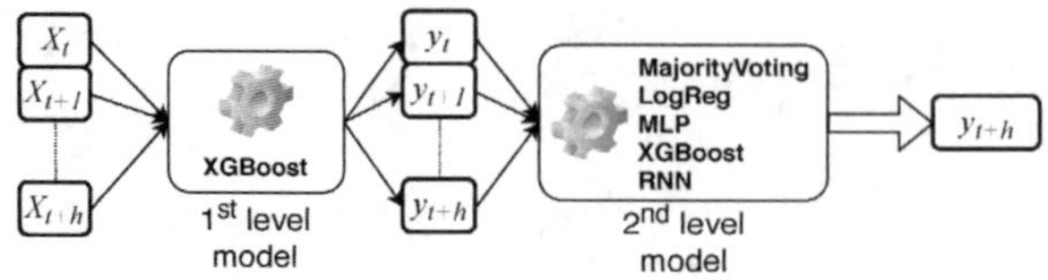

Figure 5. Structure of models with history.

Majority voting. The prediction is given by $\hat{y} = \text{Majority}(\arg\max \mathbf{p}_i)$. This can be seen as a simple baseline model. In this model each $\mathbf{p}_i$ is interpreted as an expert. The model assumes that all experts are independent and if they are not, the quality can be lower than expected. Note, that when we use the predictions of the first level models they are not independent. The model has no parameters.

Gradient Boosted Decision Trees (XGBoost). See details in section 2.3.1. In this case we used slightly different parameters: `n_estimators` $= 100$, `max_depth` $= 3$, `learning_rate` $= 0.1$.

Logistic Regression (LogReg). This is a simple linear model. The main assumption of the model is that the probability of the target label is expressed by $p(y = c|\mathbf{x}) = \frac{\exp(-\langle w_k, \mathbf{x}\rangle + b_k)}{\sum_k \exp(-\langle w_k, \mathbf{x}\rangle + b_k)}$, where $\{w_c, b_c\}_{c=1}^C, w_k \in R^d, b_k \in R$ are parameters of the algorithm, d is the dimensionality of the feature vector. We used implementation of LogReg from `scikit-learn` Python package[1].

Multi-layered perceptron (MLP) is a feed-forward neural network. The output of the model is given by $\mathbf{z}_L = a_L(W_{L-1}\mathbf{z}_{L-1} + b_{L-1}) + b_L$, where $W_l \in R^{h_l \times h_{l-1}}, b_l \in R^{h_l}$, $\mathbf{z}_0 = \mathbf{x}$ for $l = 1, \ldots, L$ and $a_l(\cdot)$ is an activation function. For $a_l(\cdot)$ we used ReLU activation function $a_l(x) = x \cdot \mathbf{1}_{x>0}$. So, MLP feeds the data to a sequence of layers where each layer performs simple non-linear mapping (activation function) of the input. Result of the composition of the mappings is highly non-linear transformation of the input data. MLP with one layer is equivalent to Logistic Regression. The parameters of the MLP, namely W_l, b_l for all l, are tuned by minimizing cross-entropy loss via Stochastic Gradient Descent (see [23]).

In an architecture of MLP that we used we set `number of layers` L to 3, as small number of layers reduce overfitting; we set `hidden_layer_sizes` h_l to $(100, 100)$, as larger hidden layer sizes provide more complex networks. In our experiments we tried different combinations of L and h_l and provide results only for the best one. For MLP we used implementation from `scikit-learn` Python library.

Recurrent neural network (RNN) is a type of artificial neural network that is applicable to sequential data due to its architecture. RNNs are called recurrent because they perform the same task for every element in a sequence, with the output being dependent on previous computations. Typically RNN looks like the one depicted in Figure 6. We used a special RNN architecture called Gated Recurrent Unit (GRU, first proposed in [24]). This model sequentially calculates the following values $z_t = \sigma(W_{zh}\mathbf{h}_{t-1} + W_{zx}\mathbf{x}_t)$, $r_t = \sigma(W_{rh}\mathbf{h}_{t-1} + W_{rx}\mathbf{x}_t)$, $\tilde{h}_t = \tanh(W_r(r_t \cdot \mathbf{h}_{t-1}) + W_x\mathbf{x}_t)$, $\mathbf{h}_t = (1 - z_t) * \mathbf{h}_{t-1} + z_t * \tilde{\mathbf{h}}_t$, where $\mathbf{h}_t$ is a hidden state and also the output of RNN which is then passed to the Logistic Regression or MLP to perform classification (so, we can think of $\mathbf{h}_t$ as features extracted from $\mathbf{x}_t$ taking into account all the previous values $\mathbf{x}_{t-1}, \mathbf{x}_{t-2}, \ldots$), $W_{zh}, W_{zx}, W_{rh}, W_{rx}, W_r, W_x$ are matrices which are tuned during

[1] https://scikit-learn.org/

training. Vector $\mathbf{h}_{t-1}$ is a feature vector for the previous element $\mathbf{x}_{t-1}$ and can be seen as a memory (past). The meaning of other values is the following

- Update gate z_t. σ in equation for update gate is a sigmoid function. So, it takes values in $(0, 1)$ interval and defines how much information to pass to the future.
- Reset gate r_t. It also takes values in $(0, 1)$ interval and defines how much the past information to forget.
- Current memory content $\tilde{\mathbf{h}}_t$. It takes values in $(-1, 1)$ interval and, thus, not a gate. It contains useful information about the $\mathbf{x}_t$ taking into account the past.
- Final memory content $\mathbf{h}_t$. It's a linear combination of the past and current memory content.

We can stack several GRU units to create deep RNN. We used the following structure of the RNN: `hidden_dim` = 128 — this is a dimension of hidden state $\mathbf{h}_t$; `n_layers` = 5 — number of layers, i.e. stacked GRU units; `bidirectional` = `True` — the same sequence can be passed to the RNN in both straight and reversed order and then the results are concatenated. This makes sense because usually RNN better memorizes the most recent elements of the sequence, but passing the sequence in both directions allows to better memorize older elements in the sequence; `n_mlp_layers` = 2, `mlp_hidden_sizes` = (64, 64) — parameters of MLP that is used on top of GRU.

Training of RNN. We used `pytorch`[2] Python library to build RNN model. RNN model is usually hard to train because we have a lot of parameters in our model and rather small data set, so it overfits easily. In order to make the procedure more stable and prevent overfitting we applied several tricks

- **Dropout**. This is a special regularization technique to reduce overfitting (see [25]). The idea is that during training random weights of the neural network are set to zero. The intuition is the following. When one set of weights is zeroed-out we have one network with smaller number of parameters, with another set of zeroed-out weights we have another network. Applying dropout is in some sense equivalent to training an ensemble of networks with smaller number of parameters (which are less prone to overfitting). We used `dropout_rate` = 0.5 — average number of weights zerod-out at each learning iteration.
- **Learning rate scheduler**. Learning rate controls how large is an update to the neural network after each step of stochastic gradient descent. We change learning rate at each iteration — start from some initial value, then monotonically decrease for several iterations, after that set it back to initial value (restart) and repeat. Such restarts allows to escape from local minimum and find more robust one with less overfitting.
- **Adding noise to the input data**. The input of the RNN is a prediction of XGBoost model which is not perfect. To take it into account and be more robust we incorporate noise to the inputs. Predictions of XGBoost is a vector of probabilities $\mathbf{p}_i$. We generate noisy probabilities by sampling them from Dirichlet Distribution with parameter $\mathbf{p}_i/\tau$, where $\tau > 0$ is a parameter. The higher its value the more noisy the samples. For values close to zero, samples almost coincide with the original vector of probabilites.

2.4. Quality assessment

The training set is rather small, therefore quality assessment is performed via **N-by-M cross-validation** procedure

(i) Split the training set into M non-intersecting random subsets.

[2] https://pytorch.org/

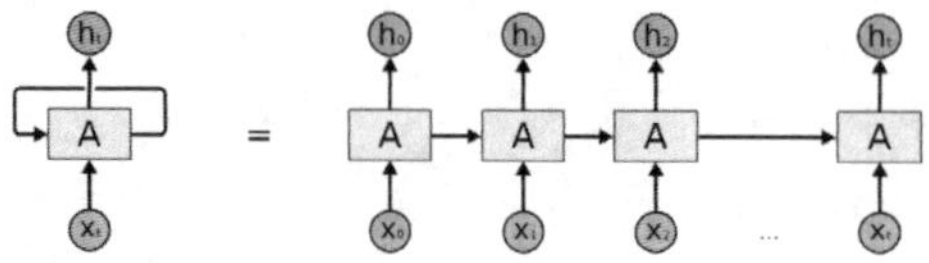

Figure 6. RNN and the unfolding in time of its forward computation.

(ii) Use $M-1$ subsets to fit the model and use the remaining one (validation set) to calculate predictions of the model

(iii) Repeat step 2 M times each time choosing different validation set. After this we obtain predictions for each point from the initial training set.

(iv) Calculate quality metrics using obtained predictions.

(v) Repeat the whole procedure N times.

(vi) Calculate mean and standard deviation of quality metrics.

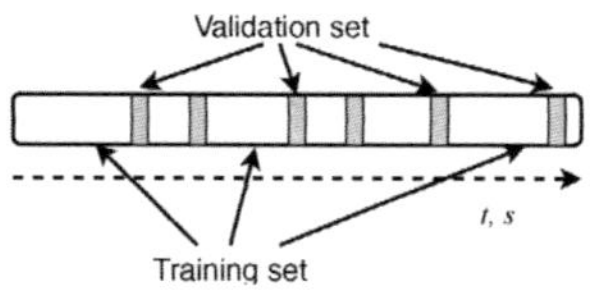

Figure 7. Splitting: without history.

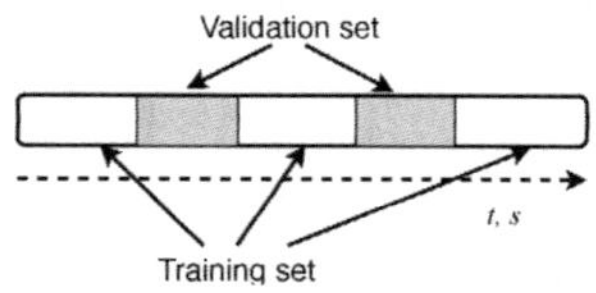

Figure 8. Splitting: with history

2.4.1. Splitting into train/validation for models with history. There is some subtlety how the data set should be splitted into training and validation folds when we use a model with history (see Figures 7 and 8).

- Splitting for model without history: validation fold is a random subset from the given data set, training fold is the rest of the data set. In this case if $\mathbf{x}_t$ is in the validation fold, then with high probability $\mathbf{x}_{t-1}, \mathbf{x}_{t+1}$ are in the training fold. So, the model just needs to interpolate one sample between $(t-1)$-th and $(t+1)$-th elements.
- Splitting for model with history: choose random $\mathbf{x}_t$ from the data set and put to the validation set the sequence $(\mathbf{x}_t, \mathbf{x}_{t-1}, \ldots, \mathbf{x}_{t-h})$ (this procedure is repeated while the desired validation fold size is not reached). The training fold is the rest of the data set. In this case the model has to interpolate large gap between $\mathbf{x}_{t-h-1}$ and $\mathbf{x}_{t+1}$.

This means that the quality of the first level models will differ for different schemes of splitting: in the first case we expect the quality to be higher than in the second case.

2.4.2. Quality metrics. The following metrics are calculated:

Accuracy. Accuracy $= \frac{1}{N}\#\{\hat{y}_i = y_i, i = 1, \ldots, N\}$, where $\hat{y}_i$ is a predicted label, y_i is a reference label and N is a dataset size. The value of Accuracy is between 0 and 1. For ideal classification Accuracy is equal to 1.

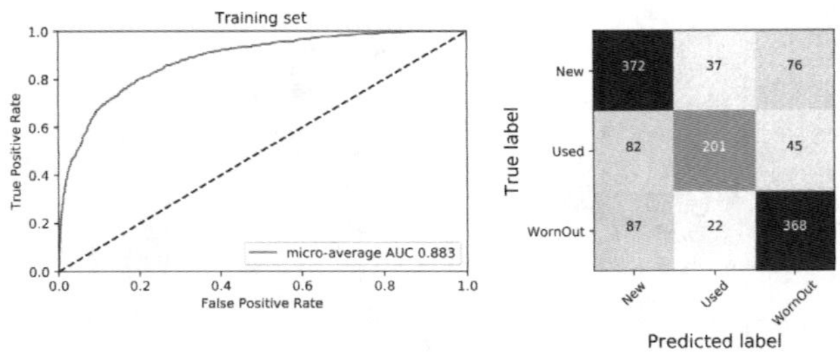

Figure 9. ROC-curve and confusion matrices for models without history: before feature selection.

F1-score for binary classification (there are two labels – positive and negative) is equal to $\mathrm{F1} = 2 * \frac{\text{precision}\cdot\text{recall}}{\text{precision}+\text{recall}}$, precision $= \frac{TP}{TP+FP}$, recall $= \frac{TP}{TP+FN}$, where TP is true positives, i.e. number correctly predicted positive labels, FP is false positives, i.e. number of incorrectly predicted positive labels, FN is false negatives, i.e. number of incorrectly predicted negative labels.

In multiclass case F1 score can be calculated for each class separately and then aggregated. In this work we used weighted aggregation which means that we take weighted sum of F1 scores for each class, weights are inversely proportional to class size. The value of $F1$-score is between 0 and 1. Good classification has $F1$-score close to 1.

Area under ROC curve. In a binary case a classificator usually returns probability p of positive class. Decision on the predicted label is made by the following rule: $\hat{y} =$ positive if $p > c$, and negative otherwise, where c is some threshold. Depending on c different TPR and FPR can be obtained ($TPR = \frac{TP}{TP+FN}$ is a true positive rate, $FPR = \frac{FP}{FP+TN}$ is a false positive rate).

ROC curve is FPR vs. TPR curve for $c \in [0, 1]$. In multiclass case we can do the following: calculate TP_k and FP_k for each class k, then compute $TP = \sum_k TP$, $FP = \sum_k FP_k$, $FN = \sum_k FN_k$, $TN = \sum_k TN$ and FPR and TPR as usual (this procedure is also called micro-averaging). ROC curve for random guess corresponds to $y = x$ line. The higher lies the curve the better works the algorithm. Qualitatively this effect is measured by calculating area under the curve (AUC).

Confusion matrix is a matrix, (i, j) element of which is a number of cases when i-th class was predicted as j-th class. For ideal classificator the matrix should be diagonal.

3. Results

In this section experimental results of application of the described approach are presented.

3.1. Models without history

In Figures 9 and 10 ROC-curve and confusion matrices are given for models before and after feature selection, the Table 1 presents the quantitative results of quality assessment. If we perform feature selection we obtain model with almost the same quality but with lower number of parameters, thus, more robust. It means that we overfit less.

Note from confusion matrix that there are brakings with `Worn Out` brakes that were classified as `New` or `Used`. The cost of such misclassification can be large. Such misclassification we will call False Negatives (FN) (in this case we suppose that positive class is `Worn Out` and all other classes are negative).

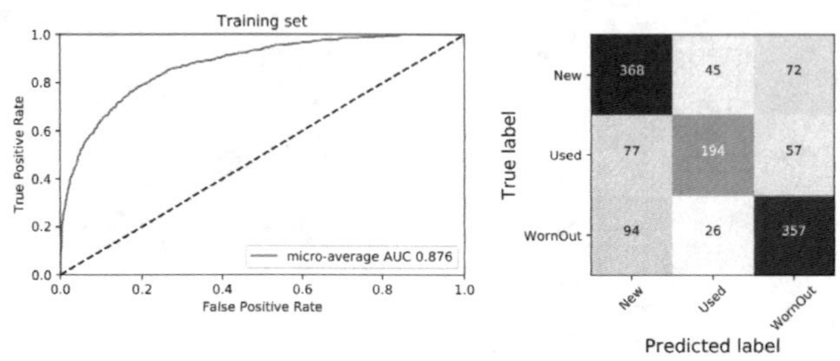

Figure 10. ROC-curve and confusion matrices for models without history: after feature selection.

Table 1. Quality metric values for models without history

Metric	Before feature selection	After feature selection
ROC AUC	0.86 ± 0.008	0.86 ± 0.009
F1-score	0.70 ± 0.019	0.70 ± 0.014
Accuracy	0.70 ± 0.019	0.71 ± 0.014

Table 2. Accuracy of models with history for history size 10

Model	Accuracy	Accuracy of 1st level model
Majority Voting	0.76 ± 0.037 (theoretical 0.94)	
LogReg	$\mathbf{0.85 \pm 0.036}$	
XGBoost	0.84 ± 0.032	0.69 ± 0.020
MLP	0.80 ± 0.043	
RNN	$\mathbf{0.85 \pm 0.033}$	

Besides XGBoost approach we also used models based on convolutional and recurrent neural networks, and support vector machine classification. However, these models provided significantly lowere accuracy.

3.2. Models with history

Table 2 provides quality of different models with history for history size 10 (here we report only accuracy as some models don't provide probabilities as predictions, only labels). We also report the accuracy of 1st level model on top of which the 2nd level model was built. For MajorityVoting we also provide the theoretical estimate of the quality assuming that 1st level models predictions are independent. Theoretical estimate can be computed as follows. Suppose that the accuracy (the probability of correct prediction) of 1st level model is p. Then, the probability that the majority of predictions is correct is equal to $P_h = \sum_{i=\lceil h/2 \rceil}^{h} \binom{h}{i} p^i (1-p)^{h-i}$.

For top-2 best performing models we conducted experiments with various history sizes (results are given in Table 3). From the table it can be seen that LogReg models accuracy drops with larger history size, while RNN's remains at the same level. Decrease in accuracy is explained by the decrease in accuracy of the first level model. For history size 10 the accuracy of LogReg and RNN models is the same, so, in this case LogReg is preferred as it is much more simple. However, the accuracy of LogReg model largely depends on the accuracy of the first level model, while RNN is more robust to it.

Table 3. Accuracy for different history sizes

Model	History size	Accuracy	Accuracy of 1st level model
RNN	10	0.85 ± 0.033	0.67 ± 0.063
LogReg	10	0.85 ± 0.036	
RNN	20	0.84 ± 0.060	0.63 ± 0.072
LogReg	20	0.82 ± 0.086	
RNN	30	0.81 ± 0.100	0.52 ± 0.140
LogReg	30	0.77 ± 0.120	
RNN	50	0.82 ± 0.145	0.57 ± 0.124
LogReg	50	0.76 ± 0.122	

Table 4. Characteristics of the second level models. Here d_{in} is dimensionality of input $\mathbf{x}_t$, n_l is total number of layers in neural network, n_t is a number of trees in XGBoost, d is a tree depth, d_h is max dimensionality of hidden layers in neural network, h is history size. Measurements were recorded on Intel(R) Core(TM) i7-7820X CPU @ 3.60GHz. Concrete values of parameters see in Section 2.3.2.

Model	# parameters	Computational complexity	Evaluation time	Space on disk
Logistic Regression	$\mathcal{O}(d_{in})$	$\mathcal{O}(hd_{in})$	0.4 ms	41 Kb
MLP	$\mathcal{O}(n_l \times d_h \max(d_h, d_{in}))$	$\mathcal{O}(hn_l d_h \max(d_h, d_{in}))$	2.4 ms	110 Kb
XGBoost	$\mathcal{O}(n_t \times (2^{d+1}))$	$\mathcal{O}(hn_t d)$	0.33 ms	815 Kb
RNN	$\mathcal{O}(n_l \times d_h \max(d_h, d_{in}))$	$\mathcal{O}(hn_l d_h \max(d_h, d_{in}))$	0.07 s	5 Mb

3.3. Computational and Memory Complexity

Average braking event from the given training set requires 1980 Kb. All brakings have different length, so actual value can be larger or smaller. After feature extraction 4.5 Kb memory is required to store one braking event. This value is constant for all braking events.

Characteristics of different models are given in Table 4. The most lightweight model is LogReg. Moreover, this model provides one of the best accuracies with history size 10. Therefore, this model can be used for on-board implementation. RNN model is the heaviest one, although it provides quality comparable to LogReg. However, we expect that this model will work better for larger training set. This model can be used for cloud implementation.

4. Conclusion

We considered the problem of predicting brake pad usage condition using the characteristics measured during braking. The optimal model (optimal in terms of trade-off between accuracy and the memory required to store the model) that we have constructed is based on Gradient Boosted Decision Trees (XGBoost) with preprocessing and feature extraction (accuracy ~ 0.70).

Accuracy can be improved by using history of braking events (accuracy ~ 0.85). We developed two models, that shows the same quality for a given data set. The first one is lightweight and is based on XGBoost + Logistic Regression. This model can implemented on-board. The second model is more complex and is based on XGBoost + Deep Recurrent Neural Network. It can be implemented in a cloud. We expect that this model will provide higher accuracy if trained on larger data set. We also tried several other models, though they didn't show good enough accuracy.

Acknowledgments

Evgeny Burnaev would like to thank Efim Boeru for a fruitfull collaboration in the framework of this project, and Yermek Kapushev for performing computational experiments.

References

[1] Hoffmann A, Lehner S, Weissenbacher C, Zechlin H 2018 Audi A3 e-tron Boxberg Driving Test *Technical report* Robert Bosch GmbH, CC-AS/EYB3, 74232 Abstatt, Germany.

[2] Jegadeeshwaran R, Sugumaran V 2015 Brake fault diagnosis using clonal selection classification algorithm (csca) — a statistical learning approach *Engineering Science and Technology, an International Journal* **18** 1 pp 14–23

[3] Chetan C Harlapur, Priyatamkumar Kadiyala, and Ramakrishna S 2019 Brake pad wear detection using machine learning *International Journal of Advance Research, Ideas and Innovations in Technology* pp 498–501

[4] Burnaev E 2019 Algorithmic foundations of predictive analytics in industrial engineering design *Journal of communications technology and electronics* **64** 12 pp 1485–92

[5] Burnaev E 2019 On construction of early warning systems for predictive maintenance in aerospace industry *Journal of communications technology and electronics* **64** 12 pp 1473–84

[6] Burnaev E 2019 Rare failure prediction via event matching for aerospace applications *In Proceedings of the 3rd International Conference on Circuits, System and Simulation (ICCSS-2019)* pp 214–20

[7] Smolyakov D, Sviridenko N, Burikov E, and Burnaev E 2018 Anomaly pattern recognition with privileged information for sensor fault detection *In Proc. ANNPR* vol 11081 (Springer LNCS Proceedings) pp 320–32

[8] Burnaev E, Erofeev P, and Papanov A 2015 Influence of resampling on accuracy of imbalanced classification *In Proc. SPIE* vol 9875

[9] Burnaev E, Erofeev P, and Smolyakov D 2015 Model selection for anomaly detection *In Proc. SPIE* vol 9875

[10] Burnaev E and Smolyakov D 2016 One-class svm with privileged information and its application to malware detection *In Proc. IEEE 16th ICDM Workshops* pp 273–80

[11] Smolyakov D, Korotin A, Erofeev P, Papanov A, and Burnaev E 2019 Meta-learning for resampling recommendation systems *In Proc. SPIE* vol 11041

[12] Burnaev E and Golubev G 2017 On one problem in multichannel signal detection *Problems of Information Transmission* **53** 4 pp 368–80

[13] Artemov A and Burnaev E 2015 Ensembles of detectors for online detection of transient changes *In Proc. SPIE* vol 9875

[14] Artemov A, Burnaev E, and Lokot A 2015 Nonparametric decomposition of quasi-periodic time series for change-point detection *In Proc. SPIE* vol 9875

[15] Volkhonskiy D, Burnaev E, Nouretdinov I, Gammerman A, and Vovk V 2017 Inductive conformal martingales for change-point detection *In Proc. of the 6th Workshop on Conformal and Probabilistic Prediction and Applications* vol 60 (Proceedings of Machine Learning Research) pp 132–53

[16] Safin A and Burnaev E 2017 Conformal kernel expected similarity for anomaly detection in time-series data *Advances in Systems Science and Applications* **17** 3 pp 22–33

[17] Ishimtsev V, Bernstein A, Burnaev E, and Nazarov I 2017 Conformal k-nn anomaly detector for univariate data streams *In Proc. of the 6th Workshop on Conformal and Probabilistic Prediction and Applications* vol 60 (Proceedings of Machine Learning Research) pp 213–27

[18] Smolyakov D, Sviridenko N, Ishimtsev V, Burikov E, and Burnaev E 2019 Learning ensembles of anomaly detectors on synthetic data *In Proc. ISNN: Advances in Neural Networks* pp 292–306

[19] Burnaev E, Koptelov I, Novikov G, and Khanipov T 2017 Automatic construction of a recurrent neural network based classifier for vehicle passage detection *In Proc. SPIE* vol 10341

[20] Wang Z, Yan W, and Oates T 2017 Time series classification from scratch with deep neural networks: A strong baseline *In Proc. International Joint Conference on Neural Networks (IJCNN)* pp 1578–85

[21] Burnaev E and Menshikov I 2009 A model of the functional state of participants of laboratory markets *Journal of Computer and Systems Sciences International* **48** 1002

[22] Chen T and Guestrin C 2016 XGBoost: A scalable tree boosting system *In Proc. of the 22nd ACM SIGKDD, KDD'16* pp 785–94

[23] Bishop C 2006 *Pattern recognition and machine learning* (Springer)

[24] Cho K, Merrienboer B, Gulcehre C, Bahdanau D, Bougares F, Schwenk H, and Bengio Y 2014 Learning phrase representations using rnn encoder-decoder for statistical machine translation *arXiv preprint* arXiv:1406.1078

[25] Srivastava N, Hinton G, Krizhevsky A, Sutskever I, and Salakhutdinov R 2014 Dropout: a simple way to prevent neural networks from overfitting *The journal of machine learning research* **15** 1 pp 1929–58

IOP Conf. Series: Materials Science and Engineering **904** (2020) 012013 doi:10.1088/1757-899X/904/1/012013

High-precision analytical TDoA positioning algorithm for eliminating the ambiguity of coordinates determination

V D Kuptsov[1], S I Ivanov[1], A A Fedotov[1] and V L Badenko[1]

[1]Peter the Great St. Petersburg Polytechnic University, 29 Polytechnicheskaya str., Saint Petersburg 195251, Russian Federation

E-mail: vdkuptsov@yandex.ru

Abstract. An analytical method is proposed for determining the position of an object based on the analysis of Time Difference of Arrival (TDoA) of signals from the transmitter of the object to the base stations (BS) receivers, which eliminates the ambiguity in determining the coordinates of the object on the plane and improves the accuracy of determining coordinates when the TDoA measurements are out of sync with base stations.

1. Introduction

Accurate remote determination of the coordinates of objects on a plane and in space is necessary in many technical applications, such as driving unmanned vehicles of the "smart city", robotic production complexes, security systems in the city and in the workplace. The coordinates of the object are determined by receiving a radio signal transmitted by the object by several base stations. In global positioning systems (for example, GPS), the radio signal is received by navigation satellites; it is possible to use specialized microwave radars within the city or a production facility. The main task of positioning the object is to determine the coordinates (position) of the object on the ground, which can be carried out using various algorithms. These include: RSSI (Received Strength Signal Indication) - the distance to the object is estimated by the signal power; AoA (Angle of Arrival) - the location of the object is determined within the triangle area formed by the intersection of the axes of the antenna patterns of the sectors of the three base stations (modified triangulation method); ToF (Time of Flight) - measuring the travel time of an electromagnetic wave from an object transmitter to a base station by using a signal with linear frequency modulation; ToA (Time of Arrival) —measurement of the signal transit time from the mobile terminal to the base station, at which the distance to the object is calculated based on the difference in the time the signal was sent and received; TDoA (Time Difference of Arrival) - measures the difference in the time of arrival of a signal from an object to several base stations; RTT (Round TripTime) - the base station sends a signal to a mobile device and waits for a response signal, the time it takes for the signal to travel in both directions is determined by the difference in the time the signal is sent and received, and therefore the distance between the objects; LPT (Location Patterning Techniques) - Location determination is performed using pattern recognition of radio signals based on sampling and recording of radio patterns of signal behavior in a specific environment.

In the difficult changing environment of the "smart city" and the production facility, the TDoA method is preferred, since this method is highly accurate and requires time synchronization only at base stations [1-5]. In the TDoA method, microwave radar measures the difference in the time of arrival of the signal from the transmitter of the object to the receivers of base stations (TDoA). The lines on which the difference in the arrival times of the microwave signal to two BSs is constant, are hyperbolas on the plane (hyperboloids in space), while both BSs are in the foci of the hyperbolas. The coordinates of the object are the intersection point of the entire set of hyperbolic curves on the plane included in the system of equations.

To obtain an estimate of the position at reasonable noise levels, the Taylor-series method [6] is also used. This is an iterative method: it starts with the initial assumption and improves the estimate at each step by defining a local linear least-squares (LS) solution. An initial assumption close to the true solution is necessary to avoid local minima. Choosing such a starting point is not easy in practice. Moreover, the convergence of the iterative process is not guaranteed. It also requires large computational resources, since LS computation is required in each iteration. In [7], it was proposed to use Taylor series expansion and linearization of hyperbolic equations. As a result, the system of equations reduces to linear in matrix form [8].

The "Time of Flight" method, using a linear frequency-modulated microwave radar waveform, is used in smart city automotive applications to automatically control the distance between moving vehicles, to warn about cross-traffic, and help in changing lanes (cross traffic alert and lane change assist), parking and detection of obstacles, pedestrians and blind spots detection (parking aid, obstacle, pedestrian and blind spots detection). In addition, radars use traffic police services to ensure traffic safety using administrative measures. Microwave range automobile radars for the simultaneous measurement of speed and range use the generation and emission of signals of a special shape, which include: frequency-modulated continuous wave (FMCW), frequency shift keying (FSK) continuous wave. A refined mathematical model of the received microwave signal, taking into account a wide range of velocity changes of the target was proposed in [9]. It is shown that the use of phase methods to determine the speed of the target significantly reduces the noise immunity of the radar, and, therefore, the minimum measurable value of the target scattering area (EPR). From this point of view, methods based on processing the amplitude spectrum of the FFT of the received radio signal are more preferable. In [10], the algorithm for estimating the speed and distance to the target as applied to the mmWave automotive radar with linear FMCW using the FFT amplitude spectrum was considered. A specific feature of signal processing algorithms based on an amplitude spectrum estimate is the occurrence of false targets when estimating the radial velocities and distances. The probability of false targets occurring increases dramatically with an increase in the total number of accompanied targets in the monitoring process. In [11, 12] we have been investigated the influence of millimeter-wave radar receiver noise on the probability of unambiguous determination of unmanned vehicles speed and range in the intelligent transportation system of the "smart city". In [13] the results of synthesis and analysis of the effectiveness of the optimal location-based system for joint detection and estimation of informative parameters of quasi-determined radar signals with frequency modulated continuous wave are presented.

In [14], an exact analytical solution of hyperbolic equations was obtained for the case when the number of TDoA measurements is equal to the number of unknown transmitter coordinates. This solution, however, cannot use TDoA measurements with additional base stations used to improve positioning accuracy. A definite drawback of the analytical method is the increased requirement for the computing power of the processor of the positioning system, since there is a need to solve nonlinear equations. However, with the development of microprocessor technology, this drawback is leveled.

The analytical method is more preferable for optimizing the calculation process and increasing the accuracy of determining the position of an object in the TDoA method. In [15], formulas are given for transforming the coordinates of a hyperbola during a shift and rotation on a plane. Based on the proposed transformations in [16], the effectiveness of two methods of TDOA analytical solution was compared to determine the possible positions of the target - intersection points of hyperboloids. The first method was based on coordinate transformation from initial system to new system, where system of equations solve is simpler, whereas the second one was based on matrix. In [17], the results of an experimental verification of an analytical method based on coordinate transformation are presented in a positioning system of three receiving antennas at a distance of 15 meters. However, the method [16] at certain positions of the object and base stations is not able to eliminate the ambiguity in determining the coordinates of the object.

The purpose of this article is to develop an analytical method for determining the position of an object based on the analysis of Time Difference of Arrival (TDoA) of microwave radar signals from the transmitter to the BS receivers, which eliminates the ambiguity in determining the coordinates of the object on the plane and improves the accuracy of determining coordinates when the TDoA measurements are out of sync with base stations.

2. The linearization method of hyperbolic equations [7]

In the absence of TDoA measurement errors and reception of direct visibility signals, the real values of the difference in the time of arrival of the TDoA signal between BS_i и BS_1 are determined by the expression

$$\Delta\tau_{i1} = \frac{r_i - r_1}{c_l} = \frac{r_{i1}}{c_l}, \quad i = 2,...M, \tag{1}$$

where M is the number of BSs, c_l is the speed of light, $r_i = \sqrt{(x_i - x)^2 + (y_i - y)^2}$ is the distance between the object and BS_i, r_1 is the distance between the object and BS_1, $[x_i, y_i]$ are the coordinates of BS_i, $[x, y]$ are the coordinates of the object. From the $r_{i1} = r_i - r_1$ follows $r_{i1} + r_1 = r_i$. Substituting the coordinates of the Cartesian system, squaring the right and left sides of the equation, expanding into a Taylor series and linearizing leads to the equation

$$(x_i - x_1)(x - x_1) + (y_i - y_1)(y - y_1) + c_l \cdot \Delta\tau_{i1} \cdot r_1 = \frac{1}{2}\left((x_i - x_1)^2 + (y_i - y_1)^2 - (c_l \cdot \Delta\tau_{i1})^2\right), \tag{2}$$

which in matrix form has the form $\boldsymbol{A\theta = b}$, where

$$A = \begin{bmatrix} x_2 - x_1 & y_2 - y_1 & c_l \cdot \Delta\tau_{21} \\ x_3 - x_1 & y_3 - y_1 & c_l \cdot \Delta\tau_{31} \\ \cdot & \cdot & \cdot \\ x_M - x_1 & y_M - y_1 & c_l \cdot \Delta\tau_{M1} \end{bmatrix} \quad b = \frac{1}{2}\begin{bmatrix} (x_2 - x_1)^2 + (y_2 - y_1)^2 - (c_l \cdot \Delta\tau_{21})^2 \\ (x_3 - x_1)^2 + (y_3 - y_1)^2 - (c_l \cdot \Delta\tau_{31})^2 \\ \cdot \quad \cdot \quad \cdot \\ (x_M - x_1)^2 + (y_M - y_1)^2 - (c_l \cdot \Delta\tau_{M1})^2 \end{bmatrix}$$

$$\theta = [x - x_1 \quad y - y_1 \quad r_1]^T \tag{3}$$

The top index $[]^T$ denotes the transpose of the matrix. Solution of the equation $\boldsymbol{A\theta = b}$ can be carried out in various ways of solving a system of linear equations. After calculating the values of the vector $\boldsymbol{\theta}$ the coordinates of the target are determined

$x = \theta_1 + x_1$; $y = \theta_2 + y_1$, where x_1 and y_1 – known coordinates of BS_1.

3. An analytical method for determining the coordinates of an object in a positioning system with 4 BS

The option of constructing a positioning system on a plane of 3 BS is the most cost-effective, but it has the disadvantage that the intersection of two hyperbolas can occur at two points on the plane. In this case, it becomes impossible to reveal the true position of the object. To eliminate the ambiguity of determining the position of the object, it is necessary to include the fourth BS in the positioning system. With the addition of the fourth BS, it becomes possible to solve three systems of equations that have one common root corresponding to the true position of the target. An analytical solution to the problem of determining the location of an object for a positioning system of 4 BSs was obtained in [16], however, not all coefficient values are given in [16], which complicates the direct use of the algorithm. In addition, the problem of eliminating the ambiguity in determining the coordinates is not resolved.

Consider a positioning system on a plane that includes four microwave signal receivers: БС$_C$, БС$_L$, БС$_R$, БС$_U$. The position coordinates of the transmitter are solutions of the system of hyperbolic equations:

$$\begin{cases} \sqrt{(x-x_L)^2+(y-y_L)^2}-\sqrt{(x-x_C)^2+(y-y_C)^2}=c\cdot\Delta\tau_{LC} \\ \sqrt{(x-x_R)^2+(y-y_R)^2}-\sqrt{(x-x_C)^2+(y-y_C)^2}=c\cdot\Delta\tau_{RC} \\ \sqrt{(x-x_U)^2+(y-y_U)^2}-\sqrt{(x-x_C)^2+(y-y_C)^2}=c\cdot\Delta\tau_{UC} \end{cases}, \text{ where} \quad (4)$$

$\Delta\tau_{LC}$, $\Delta\tau_{RC}$, $\Delta\tau_{UC}$ - the difference in the arrival times of the microwave signal between BS$_{L,R,U}$ and reference station BS$_C$.

After substituting $X = x - x_C$, $Y = y - y_C$, the system of equations can be rewritten in the form

$$\begin{cases} \sqrt{(X-X_L)^2+(Y-Y_L)^2}-\sqrt{X^2+Y^2}=L \\ \sqrt{(X-X_R)^2+(Y-Y_R)^2}-\sqrt{X^2+Y^2}=R \\ \sqrt{(X-X_U)^2+(Y-Y_U)^2}-\sqrt{X^2+Y^2}=U \end{cases}, \text{where}$$

$$\begin{aligned} X_L &= x_L - x_C,\ Y_L = y_L - y_C,\ L = c\cdot\Delta\tau_{LC} \\ X_R &= x_R - x_C,\ Y_R = y_R - y_C,\ R = c\cdot\Delta\tau_{RC} \\ X_U &= x_U - x_C,\ Y_U = y_U - y_C,\ U = c\cdot\Delta\tau_{UC} \end{aligned} \quad (5)$$

Let's note $K = \sqrt{X^2+Y^2}$, where $K > 0$, thus $K^2 = X^2+Y^2$ (6)

The system of equations (5) can be rewritten in the form

$$\begin{cases} \sqrt{(X-X_L)^2+(Y-Y_L)^2}=K+L \\ \sqrt{(X-X_R)^2+(Y-Y_R)^2}=K+R \\ \sqrt{(X-X_U)^2+(Y-Y_U)^2}=K+U \end{cases} \quad (7)$$

The squaring and reduction of the general terms leads to the form

$$\begin{cases} -2X_LX-2Y_LY=2LK+L^2-X_L^2-Y_L^2 \\ -2X_RX-2Y_RY=2RK+R^2-X_R^2-Y_R^2 \\ -2X_UX-2Y_UY=2UK+U^2-X_U^2-Y_U^2 \end{cases} \text{ or } \begin{cases} -2X_LX-2Y_LY=A+BK \\ -2X_RX-2Y_RY=C+DK \\ -2X_UX-2Y_UY=E+FK \end{cases}, \text{ where}$$

$$A=L^2-X_L^2-Y_L^2;\ B=2L;\ C=R^2-X_R^2-Y_R^2;\ D=2R;\ E=U^2-X_U^2-Y_U^2;\ F=2U \quad (8)$$

In matrix form, three systems of equations have the form:

$$\begin{bmatrix} -2X_L & -2Y_L \\ -2X_R & -2Y_R \end{bmatrix} \cdot \begin{bmatrix} X \\ Y \end{bmatrix} = \begin{bmatrix} A+BK \\ C+DK \end{bmatrix}, \quad \begin{bmatrix} -2X_R & -2Y_R \\ -2X_U & -2Y_U \end{bmatrix} \cdot \begin{bmatrix} X \\ Y \end{bmatrix} = \begin{bmatrix} C+DK \\ E+FK \end{bmatrix},$$
$$\begin{bmatrix} -2X_U & -2Y_U \\ -2X_L & -2Y_L \end{bmatrix} \cdot \begin{bmatrix} X \\ Y \end{bmatrix} = \begin{bmatrix} E+FK \\ A+BK \end{bmatrix} \tag{9}$$

Solution of the first system of equations $X = \frac{1}{\Delta_1} \cdot \begin{vmatrix} A+BK & -2Y_L \\ C+DK & -2Y_R \end{vmatrix} = M_{1X} \cdot K + N_{1X}$, где

$$M_{1X} = \frac{4}{\Delta_1} \cdot \left(RY_L - LY_R \right); \; N_{1X} = \frac{2}{\Delta_1} \cdot \left(CY_L - AY_R \right); \tag{10}$$

$Y = \frac{1}{\Delta_1} \cdot \begin{vmatrix} -2X_L & A+BK \\ -2X_R & C+DK \end{vmatrix} = M_{1Y} \cdot K + N_{1Y}$, where

$$M_{1Y} = \frac{4}{\Delta_1} \cdot \left(LX_R - RX_L \right); \; N_{1Y} = \frac{2}{\Delta_1} \cdot \left(AX_R - CX_L \right); \; \Delta_1 = 4\left(X_L Y_R - Y_L X_R \right). \tag{11}$$

Solution of the second system of equations: $X = \frac{1}{\Delta_2} \cdot \begin{vmatrix} C+DK & -2Y_R \\ E+FK & -2Y_U \end{vmatrix} = M_{2X} \cdot K + N_{2X}$, where

$$M_{2X} = \frac{4}{\Delta_2} \cdot \left(UY_R - RY_U \right); \; N_{2X} = \frac{2}{\Delta_2} \cdot \left(EY_R - CY_U \right); \tag{12}$$

$$Y = \frac{1}{\Delta_2} \cdot \begin{vmatrix} -2X_R & C+DK \\ -2X_U & E+FK \end{vmatrix} = M_{2Y} \cdot K + N_{2Y}, \quad \text{где}$$

$$M_{2Y} = \frac{4}{\Delta_2} \cdot \left(RX_U - UX_R \right); \; N_{2Y} = \frac{2}{\Delta_2} \cdot \left(CX_U - EX_R \right); \; \Delta_2 = 4\left(X_R Y_U - Y_R X_U \right). \tag{13}$$

Solution of the third system of equations: $X = \frac{1}{\Delta_3} \cdot \begin{vmatrix} E+FK & -2Y_U \\ A+BK & -2Y_L \end{vmatrix} = M_{3X} \cdot K + N_{3X}$, where

$$M_{3X} = \frac{4}{\Delta_3} \cdot \left(LY_U - UY_L \right); \; N_{3X} = \frac{2}{\Delta_3} \cdot \left(AY_U - EY_L \right) \tag{14}$$

$$Y = \frac{1}{\Delta_3} \cdot \begin{vmatrix} -2X_R & C+DK \\ -2X_U & E+FK \end{vmatrix} = M_{3Y} \cdot K + N_{3Y}, \text{ where}$$

$$M_{3Y} = \frac{4}{\Delta_3} \cdot \left(UX_L - LX_U \right); \; N_{3Y} = \frac{2}{\Delta_3} \cdot \left(EX_L - AX_U \right); \; \Delta_3 = 4\left(X_U Y_L - Y_U X_L \right). \tag{15}$$

Substitution (10 ÷ 15) in (6) defines a quadratic equation with respect to the variable K $a_p K^2 + b_p K + c_p = 0$, where p corresponds to the choice of a pair of hyperbolas, p=1, 2, 3,

$$a_p = M_{pX}^2 + M_{pY}^2 - 1; \; b_p = 2\left(M_{pX} N_{pX} + M_{pY} N_{pY} \right); \; c_p = N_{pX}^2 + N_{pY}^2 \tag{16}$$

The roots of the quadratic equation take values

$$K_{p1} = \frac{-b_p + \sqrt{b_p^2 - 4a_p c_p}}{2a_p}, \quad K_{p2} = \frac{-b_p - \sqrt{b_p^2 - 4a_p c_p}}{2a_p}, \quad \text{moreover } b_p^2 - 4a_p c_p \geq 0 \tag{17}$$

In the event that one of the two roots K_{p1} and K_{p2} is negative for the same value of p, it can be immediately excluded from the solution, and then the other root of the equation remains in the algorithm. However, it is possible that both roots K_{p1} and K_{p2} are positive. For example, a solution for one of three systems of equations (that is, for three BS, not four) with the coordinates of the object $[x; y] = [10; 1]$ and the coordinates of BS_L $[x_L ; y_L]$=[- 40 ; 40], BS_R $[x_R ; y_R]$=[- 40 ; - 40], BS_U $[x_U ; y_U]$=[40 ; - 40] meters, determines the root values K_{p1}=63,41 and K_{p2}=54,98. In this case, it becomes impossible to determine the coordinates of the object. It is for such a fairly common case that you have to use the fourth BS in the positioning system based on the analytical method. It should be noted that the method of linearization of hyperbolic equations with the indicated combination of object coordinates and base stations generally leads to a gross error in determining the coordinates of the object.

Substitution of the roots K_{p1} and K_{p2} in equations (10 ÷ 15) defines six possible pairs of coordinates of the object.

$$[x_{p1} = X_{p1} + x_C, \ y_{p1} = Y_{p1} + y_C] \text{ и } [x_{p2} = X_{p2} + x_C, \ y_{p2} = Y_{p2} + y_C], \text{ где}$$

$$X_{p1} = M_{pX} \cdot K_{p1} + N_{pX}, \quad Y_{p1} = M_{pY} \cdot K_{p1} + N_{pY},$$

$$X_{p2} = M_{pX} \cdot K_{p2} + N_{pX}, \quad Y_{p2} = M_{pY} \cdot K_{p2} + N_{pY} \tag{18}$$

If there are no TDoA measurement errors from six pairs of calculated coordinates, the three pairs of coordinates will be the same. It is this decision that is true. To identify it, twelve distances D between the positions of the object determined by (18) are calculated:

$$D_{p1\_q1} = \sqrt{\left(x_{p1} - x_{q1}\right)^2 + \left(y_{p1} - y_{q1}\right)^2} \tag{19}$$

$$D_{p2\_q2} = \sqrt{\left(x_{p2} - x_{q2}\right)^2 + \left(y_{p2} - y_{q2}\right)^2} \tag{20}$$

where both indices p and q take the values 1, 2, 3, and $p \neq q$.

The next step is to search for the minima (19) and (20) and the corresponding minima of the indices. The values of the minima (19) and (20) are compared, the smallest is chosen. The coordinates of the object $[x; y]$ are selected from the coordinate values (18) by the index of the smallest of (19) and (20) minimum.

4. Comparison of algorithms for inaccurate TDoA measurement

TDoA measurement errors occur mainly due to time out of sync at base stations and are the main reason for the inaccuracy in determining the position of an object. Of the six pairs of coordinates of the proposed algorithm, only one pair under the conditions of noise of time synchronization at the BS is as close as possible to the true coordinates of the object. The remaining pairs are either false, or determine the coordinates with a larger error. The proposed algorithm selects precisely the most accurate pair of coordinates $[x; y]$ of the object.

Using a computer experiment in the LabVIEW environment, we investigated the dependence of the standard deviation (*RMS*) of the estimated coordinates of the object on the standard deviation of the difference in the arrival times of microwave signals in a positioning radar. For this, a Gaussian white noise with the same standard deviation value for all BSs was added to the TDoA value at the input of each receiver of the positioning system. To determine the standard deviation of the estimate of the position of the object, 100 computational experiments were carried out for each value of the standard deviation of Gaussian white noise. The standard deviation of the estimate of the coordinates of the object was determined in accordance with the expression:

$$RMS = \sqrt{\frac{1}{n}\sum_{i=1}^{n}\left[\left(x_i - x\right)^2 + \left(y_i - y\right)^2\right]}, \tag{21}$$

where index i is the number of the experiment, $[x_i ; y_i]$ - estimate of the coordinates of the object in the i - th experiment, $[x; y]$ are the true values of the coordinates of the object. In figure 1 shows the dependences of the standard deviation of the estimate of the coordinates of the object on the standard deviation of the Gaussian white noise of the difference in the arrival times of microwave signals in the BS for the matrix linearization algorithm and the proposed analytical algorithm. Dependencies are calculated for the coordinates of the object $[x; y] = [10; 1]$ and coordinates of BS_C $[x_C ; y_C]$=[- 40 ; 40], BS_L $[x_L ; y_L]$=[- 40 ; - 40], BS_R $[x_R ; y_R]$=[40 ; - 40], BS_U $[x_U ; y_U]$=[40 ; 40] meters.

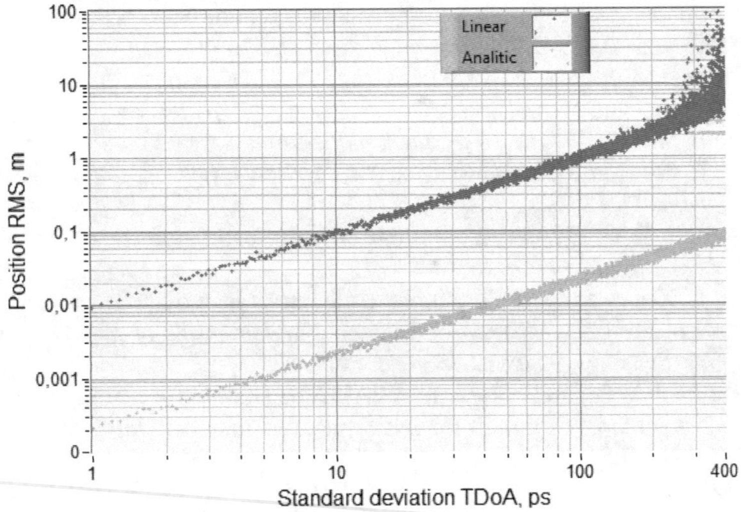

Figure 1. The dependence of the standard deviation of the estimate of the coordinates of the object from the standard deviation of the Gaussian white noise of the difference in arrival times

From the simulation results it follows that the proposed algorithm provides an accuracy of determining the coordinates of the object 50 times higher compared to the method of linearization of hyperbolic equations. The proposed method works with high accuracy (5 cm) when the BS is out of sync up to 300 ps, while the linearization method begins to produce gross errors (failures) even when the BS is out of sync in 100 ps. It can be seen from all that the proposed algorithm significantly exceeds the widely used positioning algorithm based on the linearization of hyperbolic equations.

5. Conclusion

An analytical positioning algorithm based on measuring the difference in the arrival times of signals from the transmitter of the object to the base stations eliminates the possibility of the appearance of territorial zones on the plane in which a strict solution of hyperbolic equations leads to ambiguity in determining the coordinates. The proposed algorithm has an accuracy of determining the coordinates ten times higher in comparison with the method of linearization of hyperbolic equations, less sensitive to time out of sync at base stations. Due to these advantages, the proposed algorithm is promising for remote determination of the parameters of unmanned vehicles in the technologies of the "smart city".

As future Works we propose make practice implementations to compare the performance with the simulations results. We will also expand the proposed algorithm for determining the coordinates of a target in space (3D model) and apply the method for multi-target mode.

Acknowledgments
The reported study was funded by RFBR, project number 19-29-06034.

References

[1] Al-Samahi S A, Zhang Y and Ho K C 2020 Elliptic and hyperbolic localizations using minimum measurement solutions *Signal Processing* **167** 107273

[2] Díez-González J, Álvarez R, González-Bárcena D, Sánchez-González L, Manuel Castejón-Limas M and Perez H 2019 Genetic Algorithm Approach to the 3D Node Localization in TDOA Systems *Sensors* **19** 3880

[3] Díez-González J, Álvarez R, Sánchez-González L, Fernández-Robles L, Pérez H and Castejón-Limas M 2019 3D TDOA problem solution with four receiving nodes *Sensors* **19** 2892

[4] Dunham J, Simmons J and Shapero S 2019 Analytic Emitter Geolocation and Filtering via Time Difference of Arrival *IEEE Aerospace Conference* 8741922

[5] Fokin G A, Kireev A V and Al-Odhari A H 2018 TDOA positioning accuracy performance evaluation for arc sensor configuration *Conference: 2018 Systems of Signals Generating and Processing in the Field of on Board Communications* 8350644

[6] Foy W H 1976 Position-location solutions by Taylor-series estimation *IEEE Trans. Aerosp. Electron. Syst.* AES **12** 187–194

[7] Chan Y T and Ho K C 1994 A simple and efficient estimator for hyperbolic location *IEEE Transactions on Signal Processing* **42** 1905–15

[8] Yang K, An J, Xu Z and Bu X 2008 A Generalized Total Least-Squares Algorithm for Hyperbolic Location *IEICE Transactions on Fundamentals of Electronics Communications and Computer Sciences* 91-A **12** 3824–27

[9] Ivanov S I, Kuptsov V D and Fedotov A A 2019 The signal processing algorithm of automotive FMCW radars with an extended range of speed estimation. *Journal of Physics Conference Series* **1236**(1) 012081

[10] Kuptsov V D, Ivanov S I, Fedotov A A and Badenko V L 2019 Features of Multi-target Detection Algorithm for Automotive FMCW Radar *Lecture Notes in Computer Science* **11660** 355–64

[11] Kutsov V D, Ivanov S I, Fedotov A A and Badenko V L 2020 Millimeter Wave Radar for Intelligent Transportation Systems: a Case Study of Multi-Target Problem Solution *E3S Web of Conferences* **157** 05011

[12] Kuptsov V D, Ivanov S I, Fedotov A A and Badenko V L 2020 Multi-target method for small unmanned vehicles parameters remote determination by microwave radars *Journal of Physics Conference Series* **1515** 032045

[13] Ivanov S I, Kuptsov V D, Fedotov A A and Badenko V L 2020 CFAR multi-target detection based on non-central Chi-square distribution for FMCW *Journal of Physics Conference Series* **1515** 032059

[14] Fang B X 1990 Simple Solutions for Hyperbolic and Related Position Fixes *IEEE Transactions on Aerospace and Electronic Systems* AES-**26**(5) 748–753

[15] Doan S V and Vesely J 2015 Analytical method solving system of hyperbolic equations, *25th International Conference Radioelektronika* 7129064

[16] Doan S V and Vesely J 2015 The Effectivity Comparison of TDOA Analytical Solution Methods *Proc. 16th International Radar Symposium (IRS)* 7226350

[17] Doan S V, Vesely J and Janu P 2015 The Measurement of TDOA Short Baseline International Conference on Military Technologies (ICMT) 7153697

IOP Conf. Series: Materials Science and Engineering **904** (2020) 012014 doi:10.1088/1757-899X/904/1/012014

Research dynamic motion of a landing vehicle during the final landing stage

Vsevolod V. Koryanov [1,a], **Maria Hermosilla Heras** [2]

[1] Special Machinery, Bauman Moscow State Technical University, 2nd Baumanskaya Street,5,1,105005, Moscow, Russian Federation
[2] School of Aeronautical and Space Engineering, Technical University of Madrid, Ramiro de Maeztu Street, 7, 28040, Madrid, Spain

[a] Corresponding author: vkoryanov@bmstu.ru; vkoryanov@mail.ru; maria.hermosilla.heras@alumnos.upm.es

Abstract. Landing systems for future space missions in Earth and Mars require trustable technologies capable of achieving its aim in the most accurate way possible. Using the software Universal Mechanism (developed by Bryansk State Technical University) and the classic mechanics theory on a model based on the geometry of the ExoMars vehicle, a series of different landing situations have been simulated varying the initial conditions, the approach to the surface and the surface itself. As a result, the values of the arising overloads and forces acting on the descent vehicle and its payload were calculated during the landing stage in Earth and Mars. The information collected would help understand to what extent it is crucial to know the characteristics of the landing zone (type of soil, slope, etc.) and to control the vehicle's inclination (measured from the radial axis) to provide successful landings for future missions.

1. Introduction

Decelerating landing systems require a deep analysis of the vehicle motion during the landing stage, in case the decelerating system fails due to the arduous environmental conditions and the spacecraft has to land using only the drag and lift forces produced by its body.

For instance, the landing system of the ExoMars 2020 mission relies on a main parachute (15 m wide), which decelerates the vehicle from hypersonic to subsonic speeds, and a secondary parachute (35 m wide, the largest to ever fly on Mars [1]), able to upgrade the vehicle speed to achieve a safe landing.

On May 28 of this year, the landing system for the ExoMars 2020 mission was tested unsuccessfully, since both parachutes suffered unexpected damage during the high-altitude drop test. Apparently, the fabric of the parachutes was damaged before they could even reach their maximum load. Team members made some fixes, and then performed another drop test of the 35 m wide parachute in early August. Unfortunately, the outcome of the test was disappointingly similar, since the parachute suffered damage again [2].

This study aims to approximately determine the dynamic motion of a descent vehicle without parachute deployment in both Earth's and Mars' environment. For this purpose, 180 different simulations were conducted using the modern software Universal Mechanism. Results provided the values of the arising overloads and forces acting on the descent vehicle and its passively damped payload.

2. Mathematical and physical theories applied

The vehicle and the payload are represented in Fig. 1 with their respective coordinate systems. The main body (the vehicle without the payload) has its center of mass is situated in point O, while the center of mass of the payload container is located in point P.

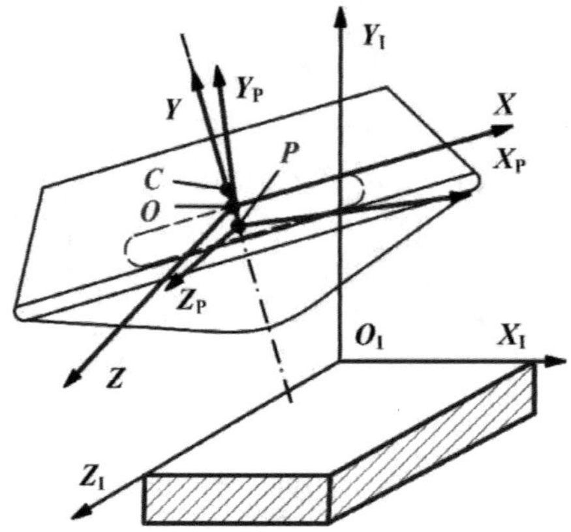

Figure 1. Systems of coordinates involved in the calculations.

The position of the payload center of mass changes with time and, consequently, the position of the center of mass of the system (point C) also changes. Assuming six degrees of freedom in the relative movement of the payload container, the movement in space of such rigid body can be described using Classical Mechanics equations. Eq.1 is the movement equation of the center of mass of the system, Eq. 2 represents its acceleration and Eq. 3 its rotational movement. Eqs. 4-5 represent, respectively, the relative and rotational movement of the payload container.

$$m\overrightarrow{a_C} = \vec{P} \quad (1)$$

$$\overrightarrow{a_C} = \overrightarrow{a_O} + \vec{\Omega} \times \overrightarrow{r_C} + \vec{\omega} \times (\vec{\omega} \times \overrightarrow{r_C}) + \overrightarrow{a_{Cr}} + 2\vec{\omega} \times \overrightarrow{v_{Cr}} \quad (2)$$

$$\frac{d\overrightarrow{K_O}}{dt} = m(\overrightarrow{v_C} \times \overrightarrow{v_O}) + \vec{M} \quad (3)$$

$$m_P\overrightarrow{a_{Pr}} = \vec{F} - m_P\overrightarrow{a_{Pe}} - m_P\overrightarrow{a_{PC}} \quad (4)$$

$$\frac{d\overrightarrow{K_P}}{dt} = \overrightarrow{M_P} \quad (5)$$

Eqs. 1-5 are projected in the system of coordinates OXYZ. They, altogether with Euler's kinematic equations (see Eqs. 6-7) and the equations that determine the transition between the system of coordinates OXYZ and $O_IX_IY_IZ_I$ (see Eq. 8), form the system of equations necessary to obtain the parameters of the descent vehicle motion when it interacts with the landing surface.

$$\left.\begin{aligned} \dot{\psi} &= (\omega_Y \cdot cos\gamma - \omega_Z \cdot sin\gamma)/cos\vartheta; \\ \dot{\vartheta} &= \omega_Y \cdot sin\gamma + \omega_Z \cdot cos\gamma; \\ \dot{\gamma} &= \omega_X - tan\vartheta(\omega_Y cos\gamma - \omega_Z sin\gamma); \end{aligned}\right\} \quad (6)$$

$$\left.\begin{aligned} \dot{\psi}_P &= (\omega_{YP} \cdot cos\gamma_P - \omega_{ZP} \cdot sin\gamma_P)/cos\vartheta_P; \\ \dot{\vartheta}_P &= \omega_{YP} \cdot sin\gamma_P + \omega_{ZP} \cdot cos\gamma_P; \\ \dot{\gamma}_P &= \omega_{XP} - tan\vartheta_P(\omega_{YP} cos\gamma_P - \omega_{ZP} sin\gamma_P); \end{aligned}\right\} \quad (7)$$

$$\left.\begin{array}{c}\dot{X}_O = v_{OX}\cdot cos\vartheta\cdot cos\psi + v_{OY}\cdot\\(-cos\gamma\cdot sin\vartheta\cdot cos\psi + sin\gamma\cdot sin\psi) +\\+v_{OZ}\cdot(sin\gamma\cdot sin\vartheta\cdot cos\psi + cos\gamma\cdot sin\psi);\\\dot{Y}_O = v_{OX}\cdot sin\vartheta + v_{OY}cos\gamma\cdot cos\vartheta +\\+v_{OZ}\cdot(-sin\gamma\cdot cos\vartheta);\\\dot{Z}_O = v_{OX}\cdot(-cos\vartheta\cdot sin\psi) + v_{OY}\cdot\\(cos\gamma\cdot sin\vartheta\cdot sin\psi + sin\gamma\cdot cos\psi) +\\+v_{OZ}\cdot(-sin\gamma\cdot sin\vartheta\cdot sin\psi + cos\gamma\cdot cos\psi);\end{array}\right\}\quad(8)$$

Where:
X_O, Y_O, Z_O are the coordinates of the point O in the system $O_IX_IY_IZ_I$;
X_P, Y_P, Z_P are the coordinates of the point P in the system OXYZ;
(v_{OX}, v_{OY}, v_{OZ}) are the projections of the vector $\overrightarrow{v_O}$ in the system OXYZ;
$(\omega_X, \omega_Y, \omega_Z)$ is the angular velocity of the vehicle without the payload;
$(\omega_{XP}, \omega_{YP}, \omega_{ZP})$ is the angular velocity of the payload;
(X_P, Y_P, Z_P) is the radio-vector of the point P;
$(v_{PrX}, v_{PrY}, v_{PrZ})$ is the relative velocity between the point P and the main body (the vehicle without the payload);
I_X, I_Y, I_Z are the moments of inertia of the main body in the system OXYZ;
I_X^P, I_Y^P, I_Z^P are the moments of inertia of the payload in the system $O_PX_PY_PZ_P$;
ψ, ϑ, γ are the yaw, pitch and roll angles of the main body, respectively;
$\psi_P, \vartheta_P, \gamma_P$ are the yaw, pitch and roll angles of the payload, respectively;

3. Landing simulations using Universal Mechanism

Universal Mechanism simulates kinematics and dynamics of planar and spatial mechanical systems, representing them as systems of rigid bodies connected by various kinematical pairs and force elements [3].

The landing is simulated with an horizontal wind velocity of 13 m/s and a vertical vehicle velocity of 35 m/s in Earth ($g = 9.81$ m/s^2) and Mars ($g = 3.73$ m/s^2) for two different types of soil: boulder and sand. Simulations are done for different body and landing surface inclinations (see Fig. 2), varying them in steps of 5°. The symmetrical axis of the landing vehicle can be deviated from the vertical axis of the planet an angle from 0° to 20°, while the landing surface can vary its inclination from -20° to 20°.

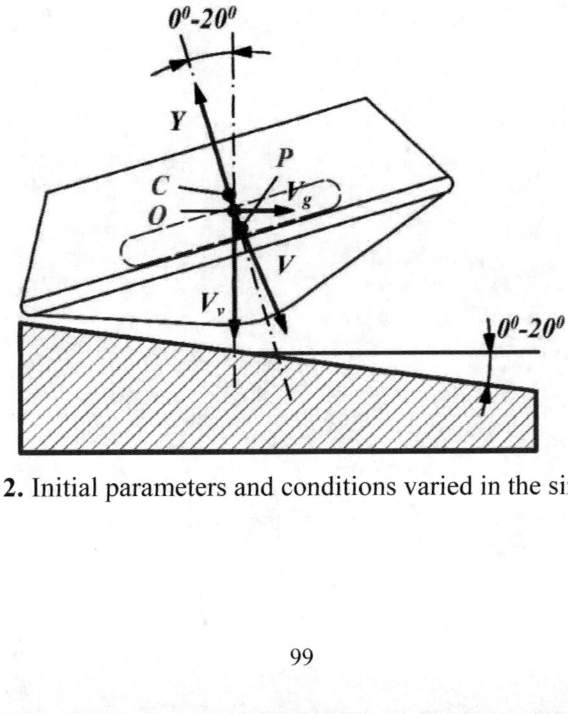

Figure 2. Initial parameters and conditions varied in the simulations.

The results obtained from the simulations consist of a set of images and data files that represent, respectively, the motion of the vehicle and the values of the arising overloads and forces acting on it and on the payload. What is important to look at when comparing these graphics is the maximum value of the overload and the time it takes to reach zero [4, 5]. However, the number of peaks that the curve might have should also be taken into account, since the material or the structure of the vehicle could be susceptible to this kind of overload behavior.

4. Comparison of the overload results on the vehicle and the payload for landings on Mars depending on the type of soil

The maximum overload value on the vehicle for landings on soft soil on Mars can be found in the 20° surface inclination simulation. Simulations with a surface inclination of -20° and -15° should not be dismissed, since a slightly second peak can be distinguished. As for the time the overload takes to fade, the fastest results can be found in the simulations of 10°, 5° and 0°.

As for landings on hard soil, the maximum overload value on the vehicle can be found in the 20° surface inclination simulation. Three peaks can be distinguished in the -20° surface inclination simulation, while for the surface inclinations -10°, -5° and 20° only two peaks can be noticed. The overload values that dissipate the fastest are found in the simulations of 15°, 10° and -15°.

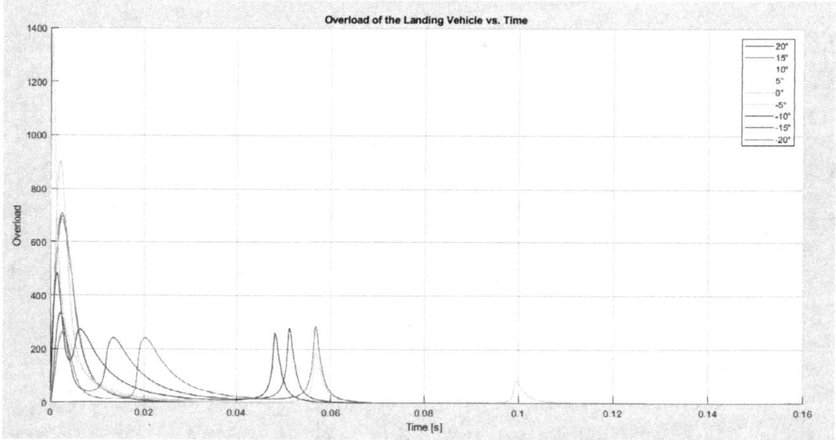

Figure 3. Overload for different landing surface inclinations acting on the landing vehicle (15° body axis inclination) for hard soil on Earth.

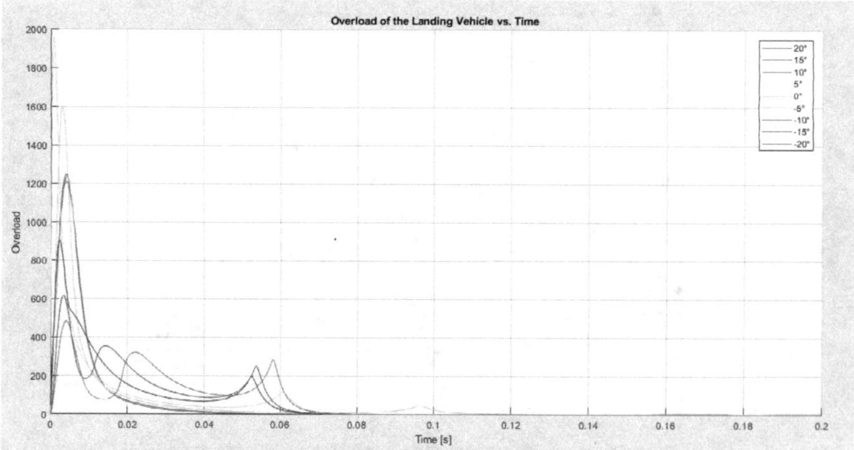

Figure 4. Overload for different landing surface inclinations acting on the landing vehicle (15° body axis inclination) for hard soil on Mars

Figs. 3-4 reveal the overload values for simulations on hard soil with a vehicle axis inclination of 15° on Earth and Mars, respectively. For simulations with a surface inclination of -20°, -15°, -10°, -5° and 0° on Mars (see Fig. 4), second and third peaks can be distinguished, but less sharp than the peaks found for the same simulations on Earth (see Fig. 3). As for the time the overload takes to fade, the fastest results can be found in the simulations of 15°, 20° and 10°. Nevertheless, the overload tends to zero faster on Earth's simulations than on Mars' simulations.

5. Conclusions

After analysing the presented data, it can be concluded that the surface inclination and type of soil have a notorious impact on the landing, as well as the deviation of the vehicle vertical axis.

Landings on hard soil develop greater values of overloads than landings on soft soil, which affect the final conditions of the payload. The landing impact is more catastrophic on hard soil surfaces, since a greater percentage of the impact energy is absorbed by the vehicle. Even though the percentage of the total overload transmitted to the payload is less in hard soil landings, the overload acting on the payload is still greater in this kind of landings than in soft soil landings. Conversely, it would be more convenient to choose the landing option that adapts more suitable to the descent vehicle structural requirements than to the payload structural requirements, since the overload acting on the payload could be reduced implementing the appropriate damping systems.

The greater the difference between the axis deviation and the surface inclination angles is, the more unstable the landing becomes. Moreover, it is more likely for second and third peaks to appear in the overload curves of surface inclination angles that differ significantly from the vehicle axis inclination angle.

The value of the overload is greater for landings on Mars than for landings on Earth. Furthermore, the third and second peaks of the overload curves are less sharp in Mars landing simulations, meaning that the overload value around those points is prolonged in time.

All in all, the optimal landing approach options that provide the safest landings would be those that occur on a slightly inclined surface made of soft soil for axis deviations as small as possible.

All the obtained results and studies can be useful in the design of new missions for the study of the planets of the solar system.

References

[1] ESA. (2019). Promising Progress for ExoMars Parachutes. Science & Exploration. https://www.esa.int/Science_Exploration/Human_and_Robotic_Exploration/Exploration/ExoMars/Promising_progress_for_ExoMars_parachutes

[2] David, L. (2019). Europe and Russia Have ExoMars Parachute Problems. It Could Threaten the 2020 Mars Launch. *SPACE.com.* https://www.space.com/exomars-rover-parachute-test-problems.html

[3] Universal Mechanism. (2019). Home Page. http://www.universalmechanism.com/en/pages/index.php?id=1#

[4] Koryanov, V. V., Kazakovtsev, V. P., Griselin, H., & Danhe, C. (2019, October). Analysis of the dynamics of movement landing vehicle with an inflatable braking device on final stage of trajectory. *IOP Conference Series: Materials Science and Engineering* (Vol. 630, No. 1, p. 012005). IOP Publishing.

[5] V.V. Koryanov, S. Huanyu. Analysis of emerging disturbing factors on the descent vehicle with inflatable mechanical devices during the descent stage. *IOP Conference Series: Materials Science and Engineering* 2019. C. 012031.

Influence of interference between the fuselage and the propeller ring on the maximum thrust of the air pusher propeller

Vladislav S. Alesin, Vitaliy V. Gubskiy, Olga V. Pavlenko

Central Aerohydrodynamic Institute (TsAGI)
[1] Zhukovsky Street, Zhukovsky, Moscow Region, 140180, Russia

Seralesin@mail.ru, vitaly.gubsky@tsagi.ru, olga.v.pavlenko@yandex.ru

Abstract. The geometrical parameters of ring profiles were presented and the estimation of its influence on the peak values of propeller thrust was described. The results of computational studies using software based on Reynolds-Averaged Navier-Stokes (RANS) equations to investigate the influence of the rear fuselage and the shape of the ring profile on the thrust of pusher propeller were presented. The pressure distribution and the velocity field changed depending on the shape of the rear fuselage and the ring profile, and their influence on the maximum thrust was shown.

1. Introduction

The propeller ring configuration is widely used in flying vehicles in order to increase thrust, to protect the propeller against damages, to increase maintenance safety and to reduce noise [1, 2]. The advantages of the propeller ring (PR) make it possible to use it for hovercraft, light aircraft, VTOL aircraft, small-sized long-range aircraft, and for solving various problems as well. The propeller ring can also serve as a fenestron tail [3] for single-rotor helicopters, low-speed aircraft and airships [4 - 6].

Nowadays the propeller ring design is used in many vehicles, and investigations of propeller ring engine (PRE) provokes significant interest [1-9]. Previous studies in this area have focused on flow features and influence of structural elements on the PRE thrust. For example, according to experimental data, which were obtained when investigating aerodynamic characteristics of disk-shaped drones with the ring propeller configuration in the T-1 MAI wind tunnel (Russia), at constant rotor speeds and increasing angle of attack, the lifting efficiency also increased [7]. It was shown that the propeller ring reduced the rotation speed of the flow in the wake of the propeller [8], and also the experiments proved that the ring profile influenced PRE performance [9].

To study the operating conditions of the propeller in the aerodynamic wake of the body of rotation, to evaluate the thrust of the propeller and the aerodynamic characteristics of the combination “body of rotation/air pusher propeller”, the experimental studies of the fuselage model with a pusher six-bladed propeller were carried out in the T-102 TsAGI wind tunnel [10]. The wind tunnel tests were conducted with a fixed number of rotor speed $N = 6000$ rpm, in the range of free-stream velocities $V = 0 \dots 50$ m/s. The maximum Reynolds number calculated from the length of the axisymmetrical body at velocity $V = 50$ m/s was $Re = 8.3 \cdot 10^6$. The diameter $D = 0.24$ m six-bladed propeller with blades made of P-107 profile, was used as a propeller.

IOP Conf. Series: Materials Science and Engineering **904** (2020) 012015 doi:10.1088/1757-899X/904/1/012015

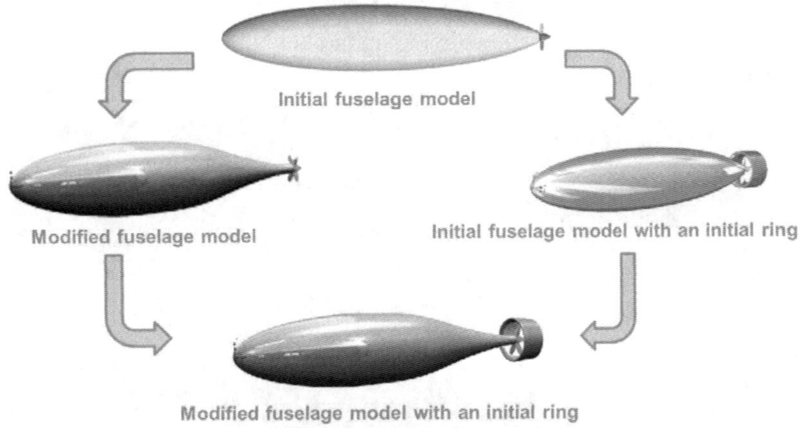

Figure 1. Computational models.

Previous numerical studies [11, 12] on increasing the thrust of an air pusher propeller by changing the rear fuselage and installing the propeller in the ring (Fig. 1) showed that:

- the use of propeller ring led to a significant increase in the thrust of the propeller (where the total thrust was determined as $T_{total} = T_{propeller} + T_{ring}$ (kg)) in the entire calculated range of flight speeds;
- the only change in the shape of the initial rear fuselage (without using a ring) of R.H. Liebeck profile [23, 14] reduced the base drag without creating a separation, and led to a slight increase in propeller thrust in the entire calculated range of flight speeds;
- the use of the ring and the change in the shape of the rear fuselage also increased the thrust, but led to a significant increase in drag with an increase in flow rate, which eliminated any benefits.

An analysis of the results of numerical studies of the initial PRE [11] with various shapes of the rear fuselage showed that with optimal regime of V = 35...40 m/s and Reynolds number $Re \approx$ 7 million, the thrust can be significantly increased without optimizing the entire system (fuselage + RPE). However, for flights with higher speeds (V > 50 m / s), it is necessary to change the shape of the ring profile due to the growth of its drag. Additionally, in order to achieve the possibility of flying at these speeds and compensate for the increasing drag due to increased thrust it is necessary to holistically optimize the outline of the rear fuselage, the profile and the angle of ring position.

In an earlier study [12], the influence of the shape of the rear fuselage on the thrust of a rotor-propulsion engine had been investigated. The results showed that a change in the shape of the rear fuselage affected the local angle of attack, at which the flow ran onto the ring, as a result of which the PRE thrust changed. In this study the investigations continued.

2. Aerodynamic design of new ring profile shape to increase PRE thrust

The aerodynamic design of a new ring shape was carried out for the same outlines of the fuselage, the unchanged chord of the ring and the fixed distance (L = 0.017 m, Fig. 2 a) from the tip of the blade to the spot where the profile of the inner side of the ring had the largest thickness.

According to the above limitations, a new ring profile was designed for numerical studies of the fuselage assembly with the PRE (Fig. 2b), which had the same chord b = 0.176 m as the initial profile, the angle of the profile of the ring is δ_{pr} = -5 deg., and the relative curvature $\bar{f}$ = 5 %, but it differed from the initial profile because of its lower relative thickness $\bar{c}_{mod.} = c / b = 14\ \%$ $(\Delta\ \bar{c} = -2\ \%)$ whereas the maximum relative thickness and curvature equally shifted backward. Moreover, for the initial profile of the ring, the position of the maximum relative thickness and curvature was $\bar{x}_{c\ ref.} = \bar{x}_{f\ ref.}$ = 23 %, and for the modified profile - $\bar{x}_{c\ dom.} = \bar{x}_{f\ mod.}$ = 53 %.

The angle of the ring aperture was selected based on the analysis of the distribution of flow rates in the area where the ring was installed so that the flow ran onto the ring at an angle corresponding to the regime of maximum lift-to-drag ratio of the ring profile.

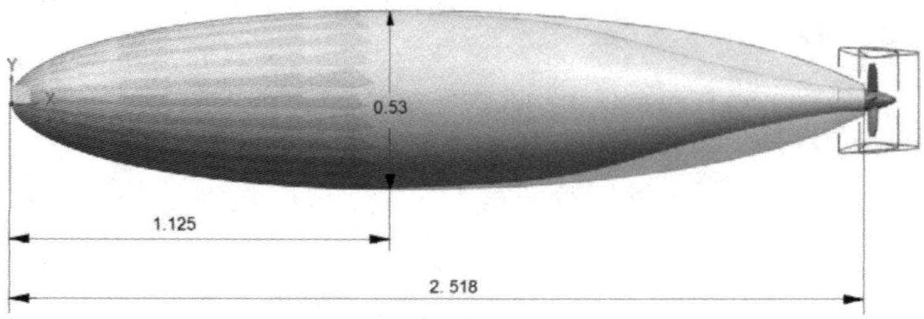

a)

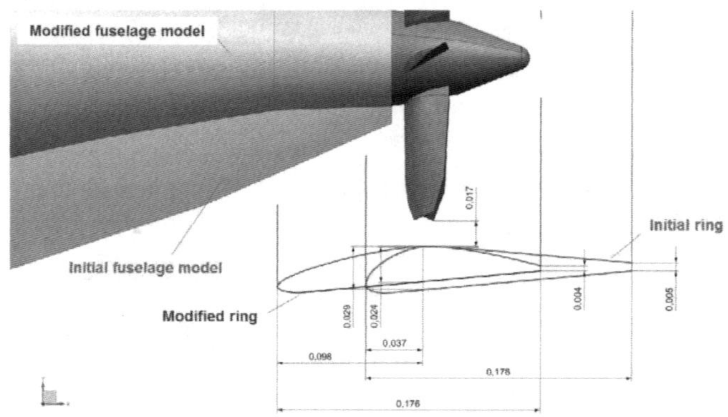

b)

Figure 2. General layout and geometrical data (m) of computational models.

To maintain the unchaged gap between the maximum thickness of the ring profile and the tip of the blade, the ring with a new (thinner) profile was shifted forward in the flow compared to the position of the initial ring (Fig. 2, b).

Earlier calculations [12] had showed that the local angle of attack of the ring profile $\alpha = 5$ deg corresponded to the maximum lift-to-drag ratio of the isolated profile. Therefore, in the calculated parameters, $\alpha = -1$ deg. corresponded to the local angle of attack of the ring profile with the modified fuselage, and $\alpha = 7$ deg. corresponded to the local angle of attack of the profile with the initial fuselage, in which the isolated profile had the greatest rarefaction on the upper surface and the maximum lift-to-drag ratio. To make calculations of the designed profile of the ring, these angles were chosen as the base.

3. Numerical studies of new PRE shape

Numerical studies of the initial and modified fuselage shapes with the initial and modified ring profiles were performed using RANS-equations on the structured mesh containing about 17 million cells (Fig. 3). For the simultaneous calculation of the translational motion of the fuselage and the rotation of the propeller, two computational domains were chosen: the first domain simulated the air flow onto the studied model, and the second domain – the rotational air motion simulating the rotation of the propeller. [15]. Due to the fact that the simulation models were bodies of rotation and the pusher propeller had 6 blades with a fixed angle the calculation grids for 1/6 of the model with areas of rotational and translational air flow and periodic boundaries were built. Finally, the full model was simulated in RANS-equations. For a more correct comparison of the calculation results, one such computational grid was adapted to the geometry of all the studied models.

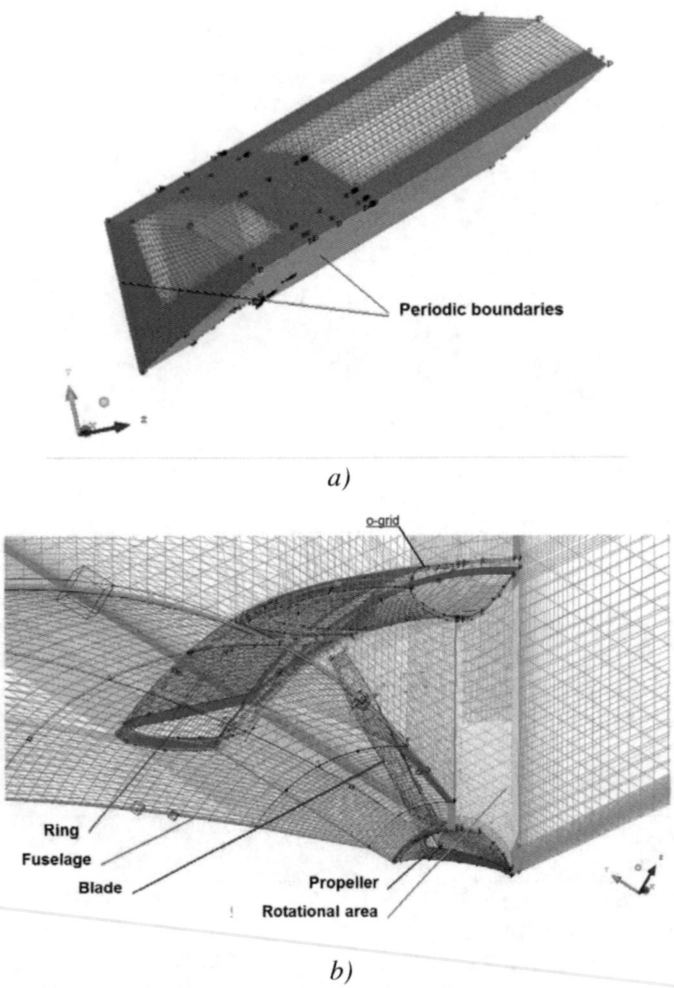

Figure 3. Computational grid 1/6 sector.

The calculations were carried out at a fixed rotor speed $N = 6000$ rpm, in the range of free-stream velocities $V = 0 \div 50$ m/s in accordance with the experimental conditions in the T-102 wind tunnel. The propeller was not modified.

It is worth noting that, according to the calculations, when the free-stream velocity was $V \leqslant 35$ m / s, the modification of the rear fuselage led to an increase in the efficiency of the propeller without a ring, however, at a speed of $V \geqslant 40$ .. 50 m / s, the propeller thrust decreased below the level of the initial fuselage (without ring), which was shown in the paper [12];

The numerical studies showed that the maximum thrust of modified propeller ring was obtained at speeds of $V = 35$ ... 40 m/s, and a further increase in speed to $V = 50$ m/s led to a decrease in thrust by $\Delta T = 0.5$ kg from the maximum (for the initial rings $\Delta T = 1$ kg) (Fig. 4).

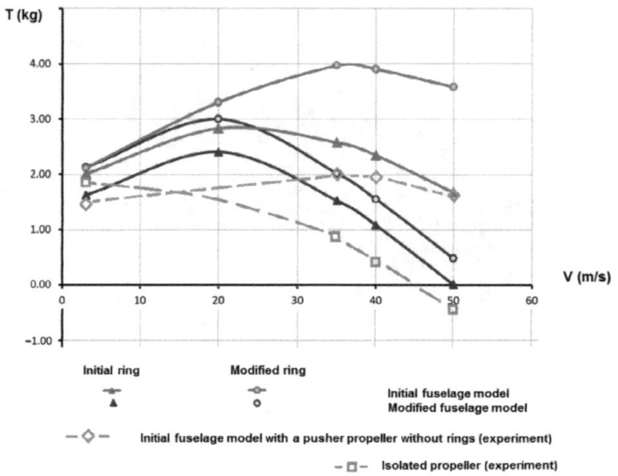

Figure 4. Dependency between the PRE thrust and the flight speed.

The PRE thrust values of the computational models, representing as the sum of propeller thrust and ring thrust, at $V = 35$ m/s, are presented in Table 1. It was shown that at a given free-stream velocity, a change in the ring profiling had a negligible effect on the thrust of the propeller itself, but had a significant effect on the thrust of the ring

Table1. PRE thrust at $V = 35$ m/s.

	T_{prop} (kg)	T_{ring} (kg)	$T_{total} = T_{prop} + T_{ring}$ (kg)
Initial fuselage model with an initial ring	1.80	0.87	2.67
Initial fuselage model with a modified ring	1.87	2.10	3.97
Modified fuselage model with an initial ring	1.64	-0.12	1.53
Modified fuselage model with a modified ring	1.62	0.40	2.02
Initial fuselage model with a pusher propeller without rings (experiment)	2.05	-	2.05
The isolated propeller (experiment)	0.93	-	0.93

The table shows that the total thrust of the PRE with a modified profile significantly exceeded the thrust of a conventional propeller (pulling and pusher) with the exception of the modified fuselage model with a modified ring. In this case, the thrust of the PRE was comparable to the thrust of the pusher propeller, however, at lower speeds (V up to 20 m/s), this solution also had advantages (see Fig. 4), which allowed us to recommend the proposed solution for low-speed aircraft.

Figures 5-7 show the pressure distribution on the surface of the simulation models and in the XOY section plane $z = 0$ at flow velocity of $V = 35$ m/s. Figure 6 shows the distribution of the pressure coefficient in the XOY section plane, $z = 0$ for the modified fuselage and the initial ring at flow velocity of $V = 35$ m/s, and Figure 7 shows the case of the initial fuselage and the modified ring. It can be seen that at the given flow velocity ($V = 35$ m/s), rarefaction was observed on almost the entire inner surface of the modified ring, and the area of greatest rarefaction occupied up to 60% of the chord of the ring profile, whereas the initial profile - only 30% (Figures 6, 7).

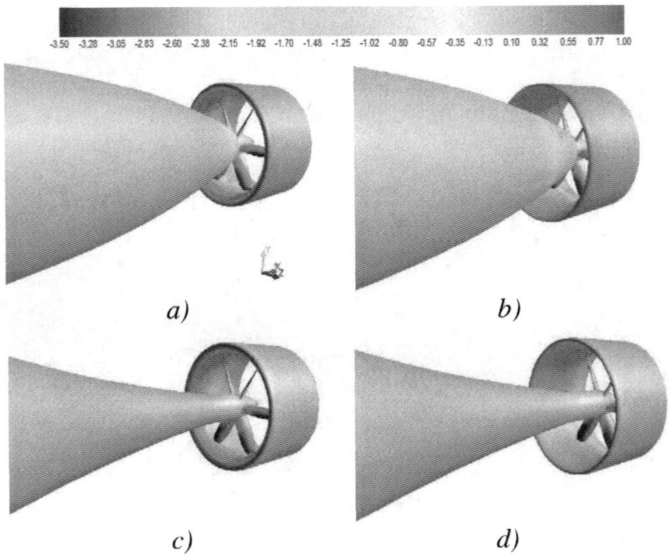

Figure 5. Pressure distribution at airflow velocity of V =35 m/s.

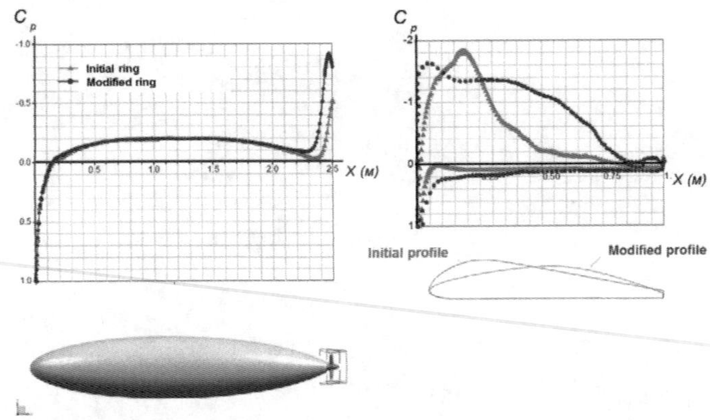

Figure 6. Pressure distribution in the XOY section plane, $z = 0$ for the initial fuselage with initial and modified rings, $V = 35$ m/s.

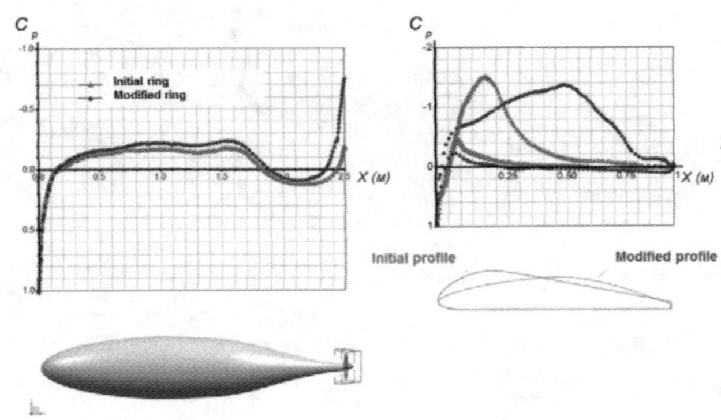

Figure 7. Pressure distribution in the XOY section plane, $z = 0$ for the modified fuselage with initial and modified ring profiles, $V = 35$ m/s.

According to the calculation results, the rarefaction on the inner surface of the modified ring took the shape of roof-top pressure distribution (Fig. 7), at which a maximum increase in the PRE thrust was observed. Whereas the dependency C_p *(x)* assumed a peak distribution (Fig. 6), an increase in the flow velocity was accompanied by a decrease in the PRE thrust.

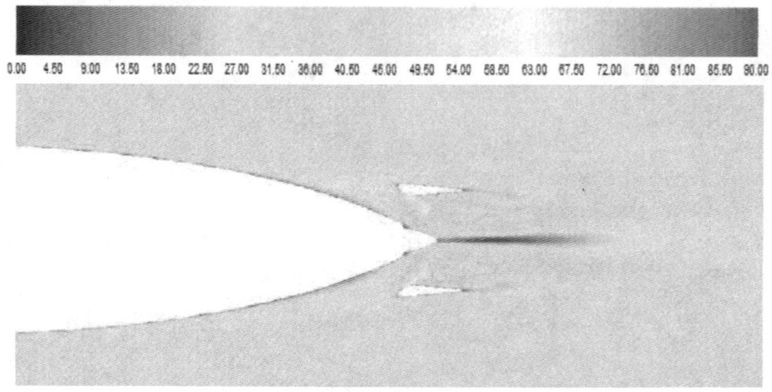

a)

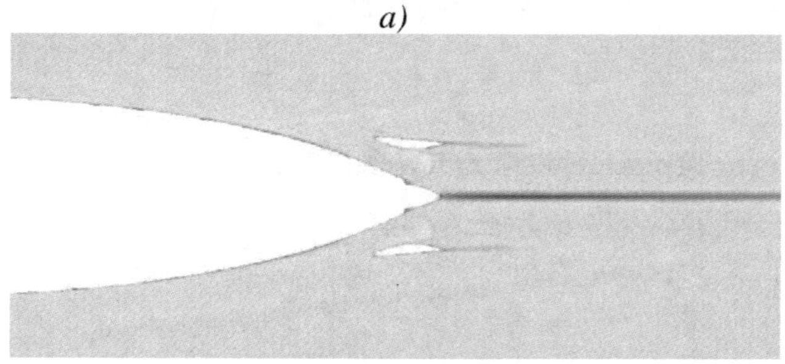

b)

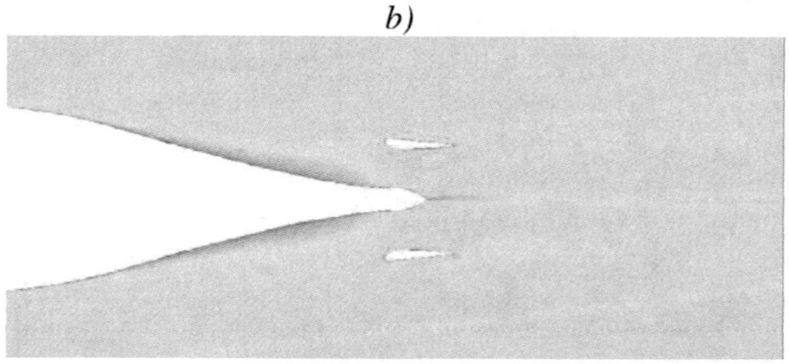

c)

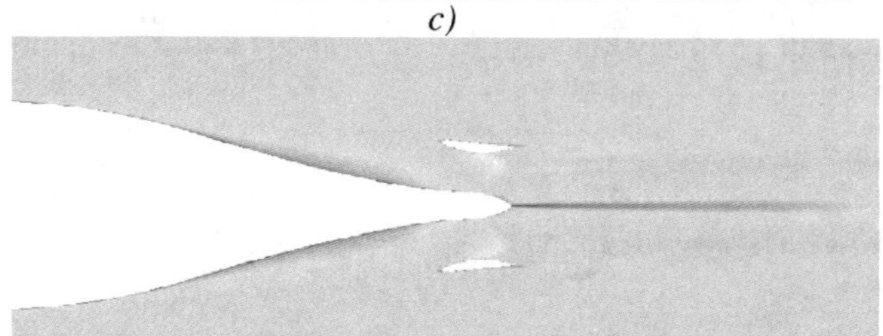

d)

Figure 8. Velocity distribution in the XOY plane, $z = 0$, $V = 35$ m/s.

The velocity distribution in the wake behind the PRE showed the direction of flow from the propeller blades at higher speeds: an even strongly inhibited wake behind the hub, convergent in the far region behind the model (Fig. 8, b).

4. Conclusion

The numerical studies showed that to design a PRE-equipped aircraft it was necessary to select parameters such as: profile thickness, chord and angle of the ring, taking into account the flow conditions and interference.

The analysis of the results of computational studies showed that at low flight speeds $V < 20$ m/s, the maximum thrust of the PRE was mainly affected by the shape of the rear fuselage, and not by the profile of the ring. The highest thrust was obtained at flow velocity of $V = 35$ m/s for the model with the initial fuselage shape and with the modified ring profile. With an increase in flight speed $V > 35$ m/s, it was necessary to take into account the increasing drag of the PRE and the operating conditions of the propeller in a thin boundary layer.

Reducing the thickness of the boundary layer and increasing the velocity in it due to a change in the shape of the fuselage allowed reducing the drag of the fuselage itself, but led to more significant loss of PRE thrust, limiting the range of applicable speeds.

The analysis of the pressure distribution data on the surface of the ring showed that the rarefaction on the inner surface of the ring, which takes the shape of a roof-top pressure distribution, corresponded to the maximum thrust of the propeller. If the dependence $C_p\ (x)$ assumed a peak distribution, then, with an increase in the flow velocity, the PRE thrust decreased. Therefore, the assumption can be made that designing a propeller ring for a certain flight speed requires the solution of the inverse problem of obtaining a roof-top pressure distribution on the inner surface of the ring.

References

[1] Abalakin I.V., Anikin V.A., Bahvalov P.A., Bobkov V.G., Kozubskaja T.K., Numerical investigation of the aerodynamic and acoustical properties of propeller in a ring. *Fluid Dynamics*, 2016, vol. 51, no. 3, pp. 419-433

[2] Skvortsov R.A., Titarev V.A., Belyaev I.V., Numerical research of noise of a propeller in a ring. *Akustika sredy obitaniya*, 2016, pp. 152–155. [in Russian]

[3] Shydakov V.I., Zavalov O.A., Numerical modeling for parametrical study of fenestron geometry on the basis of energy analysis. *Aerospace MAI Journal*, 2013, vol. 20, no. 5, pp. 7-16.

[4] Ostrouhov S.P. Aerodynamics of propellers and в Ring-Propeller Engine, Moscow, Fizmatlit, 2014, pp. 17-18. [in Russian]

[5] Shajdakov V.I., Aerodynamic characteristics of system «the Propeller in a ring» in the conditions of flow by a horizontal flow at zero angle of attack, *Nauchnyi vestnik MGTU GA*, 2016, no. 226(4), pp. 165-174.

[6] Shaidakov V.I., Zavalov O.A. *Trudy MAI*, 2011, no. 49. URL: http://trudymai.ru/eng/published.php?ID=26563

[7] Moizykh E.I., Zavalov O.A., Kuznetsov A.V., Experimental researches of aerodynamic characteristics of a distans piloted flight vehicle with a system “a propeller in a ring” *Trudy MAI*, 2012, no. 50. URL: http://trudymai.ru/eng/published.php?ID=26557

[8] Osgar John Ohanian III, Ducted Fan Aerodynamics and Modeling, with Applications of Steady and Synthetic Jet Flow Control, *Dissertation submitted to the faculty of the Virginia Polytechnic Institute and State University in partial fulfillment of the requirements for the degree of Doctor of Philosophy in Mechanical Engineering*, 2011

[9] Serdar Yılmaz1, Duygu Erdem, Effects of Duct Shape on a Ducted Propeller Performance, *51st AIAA Aerospace Sciences Meeting Including the New Horizons Forum and Aerospace Exposition*, Grapevine (Dallas/Ft. Worth Region), Texas, 07-10, 2017.

[10] Razov A.A., The Numerical analysis of propeller arrangement efficiency in a viscous layer by means of the Navier-Stokes equations *Uchenye zapiski TsAGI*, vol. XL, no. 3, 2009, pp. 28–35 [in Russian]

[11] Alesin V.S., Gubsky V.V., Druzhinin O.V., Eryomin V.J., Pavlenko O.V., *Avtomatizatsiya. Sovremennye tekhnologii*, vol. 72, no. 2, 2018, pp. 91–96. [in Russian]

[12] Alesin V.S., Gubsky V.V., Druzhinin O.V., Eryomin V.J., Pavlenko O.V., Numerical researches of flow around a rotation body and a pusher propeller in a ring // *TVF Journal*, Volume XCII, № 2, (731), 2018, pp. 10–15. [in Russian]

[13] Liebeck R.H., Low Reynolds number airfoil design at the Douglas aircraft company, *Proceedings of conference on Aerodynamics at low Reynolds numbers*, 1986. Vol. 1, paper N 7.

[14] Stratford B.S., An experimental flow with zero skin friction throughout its region of pressure rise, *J. of Fluid Mech.* , 1959. Vol. 5, N 1.

[15] Vozhdaev V.V., Teperin L.L., Tchernyshev S.L., Application and features of modern methods of flight vehicles aerodynamic characteristics calculation on the basis of solutions of the Navier-Stokes equations. *Trudy TsAGI*, issue 2740, Moscow, TsAGI im. prof. N.E. Zhukovskogo, 2014, 62 p [in Russian]

AUTHOR INDEX

Institute of Physics
Dirac House, Temple Back
Bristol BS1 6BE UK

ISSN: 1757-8981
ISBN 978-1-7138-2814-3

CURRAN ASSOCIATES INC.
proceedings.com